Andrew McKnight
Treasures in My Chest

www.spiritninjallc.com

Published by Spirit Ninja LLC
First Edition print
www.spiritninjallc.com

Cover art and design: Stilson Greene
Author Photos: Christi Porter Photography
Copy Editor and Proofreader: Chris Nicholson
Content Reviewer: Maria de los Angeles
Maps: Howard Mathieson

Treasures in My Chest
ISBN 978-0-578-63083-0
Printed and distributed by Ingram Spark

Editorial inquiries info@andrewmcknight.net,
www.AndrewMcKnight.net

Treasures in My Chest

Andrew McKnight

This book has a companion music album. If one is not included, you are entitled to download it for your personal use and enjoyment - please email info@andrewmcknight.net for download instructions.

Front Cover Images:
Top row: Great grandpa Andrew the Fiddler plays for young Jane at his 85th and final birthday; his wife Maggie Jane Robinson (1894); Gram with my dad about 1945; page 1 of the song "Margaret" (cover page also shown)

Middle: Mom's grand uncle Roy Baker, killed in France Aug. 4, 1917; Mom's father on his grandfather David Bagley's lap (abt. 1910), Aretas Culver in Civil War uniform; Gramps circa 1926.

Bottom row: Gramps in the baby basket, flanked by sisters Margaret and Sarah; mom's parents; Gram with her parents; the wedding locket of Gram's paternal grandparents Julia and Francis; Gramp's sister Margaret.

Dedication

To my grandmother Madeleine, who started me on this journey as a barely interested kid, kept and shared our stories with reverence and enthusiasm, and whose spirit is with me in many ways every day,

And to my daughter who wears her name, to my wife who brought her into the world, and to my mother who delivered me.

Indeed I am held in a circle of four strong women.

Bless them all from their near and distant times on this earth, the rogues and the renegades, the saints and the Samaritans. Their sins are no more ours to bear than their accomplishments are for our credit. They swim eternally in our veins, and we each need each and all of them to be here, magical, mysterious, miraculous. Give thanks for the complicated mosaic of identities we inherited from them, as well as their genetic assets and liabilities coiled and woven into every one of our cells.

It is their gift of your life and mine, which we gladly and gratefully acknowledge.

Table of Contents

Foreword

The hesitation to be edited - and an attachment to our words - comes from not trusting that there's more where that came from, and assuming (falsely) that our words are objects that belong to us (apart from copyrights, of course). But in the end, writing belongs to no one - it only comes through you.

To borrow from the literary world: the words are forged in the "smithy of my soul" (James Joyce), but "your children are not your children" (Kahil Gibran), and there is "one story and one story only" (Robert Graves) ... and yet we are all meaning-makers who come alive to tell a story. Everything else is merely technique.

Coming alive to tell a story ...

Treasures in My Chest weaves a beautiful tale of things that are of us but ultimately do not belong to us, that are passed on, or meet dead ends. By exploring his ancestry through the molecular language of DNA, Andrew offers a perfect example of origin story that comes through us, and our personality, but is not rigidly fixed. In his family, future generations will continue to unravel the threads of a really good yarn about Irish immigrants, the Civil War, cross-Atlantic connections, seafaring, and a grown man who journeys the uncharted territories of his inner child.

As a singer and songwriter, Andrew never thought to pen a full-length book about his curiosity for genealogy, but I encouraged him to see that the form follows the function. I was delighted by poetic passages among the more straightforward details about scouring for family records, which certainly piqued my curiosity about my own ancestry. The Irish and Cuban immigration experience in the U.S. have key similarities!

Learning about the detective work involved in exploring the branches of a family tree circles me back to the act of editing. In a way, even just exploring your background is an act of being "edited." We surrender to the story of our ancestors, and let it unfold, and mold it to serve our better angels, but always in light of truth that is deeply personal. Who's in your life to help you be your best self? Sometimes it might be a voice from the dead.

The book and his original songs on the album also work organically together, and offer a glance into every artist's creative process, which in Andrew's case, involved poignant recordings with both his daughter and father.

So my editor's eye and teacher's heart revealed once more the magic of making meaning where stories remain untold. His project is delicious food for thought about writing and legacy, which resonates with my own writing about becoming a parent to my Alzheimer's parents. I never gave birth, so for me, my words are my legacy to the world -- a way of preserving memory. We do, as writers, most certainly give birth to words, but my father's particular genetic code will no longer be embodied.

When we write a story about "us" through "me," knowing it's just passing through, some other story recombines -- like DNA -- into the collective consciousness. Above all, Andrew's book rings true in refreshing awareness about coming alive to tell a story, and may inspire you to tell your own!

Maria de Los Angeles,
Washington, D.C.
January 2020

Preface

Writing this book acknowledges that I won't always be here to tell my story and that my stories are the product of literally millions of other stories that came before mine. No problem, right?

Since I started poking around some seven years ago, I have been exceedingly lucky to uncover much of my family's history from the last couple hundred years, and for that I am deeply grateful. But this project full of stories and songs probably would not have happened had it not been for a chance discovery in the summer of 2016, when I learned that my great-grandfather was a songwriter. I knew that I had inherited his name. But when I held the three pages of his song "Margaret," published in 1906, I wondered about his contribution to my musical inheritance for the first time.

While my stories are unique to me, we all have them, even if we don't know much of anything about the stories or the people who lived them. Some are tragic, some are noble - and most are somewhere in between. People went about their everyday lives in the manner of their time period, struggling to make ends meet, trying to feed their families and raise their children, and coping with uncertainty and change - just like their descendants. In our genes are wound the best of their genetic gifts and talents as well as their limits and liabilities. I gradually came to see my journey as simply a realization of the potential in each of us bestowed upon us by the collective potential of our ancestors. And perhaps as with many things human, that potential adds up to more than the sum of the parts.

What I did not anticipate has been the greatest treasure of all - connecting with other people who share pieces of those stories. Some of them had different pieces of the same puzzle, while for others our connection and history were essentially completely new discoveries. There is a bonding in this collaborating as we get to know one another, usually via long distance electronic text of some fashion. In sharing some of our family experiences, we might discover traits or hand-me-downs that come from our common family. I certainly was interested in exploring my musical inheritance, but I was rewarded with so much more.

This book is not only about the stories that resulted from that research, but rather the human journey of connection and community. I realized as I began writing that we are all inherently curious about who we are, where we came from and why we are here. I began to see my experiences simply as an interesting lens to give others an appreciation of their inherited gifts - the known and the unknown - and perhaps inspire their curiosity and empower them to dig deeper.

I never really imagined that my ancestry would wind up braided together with my art. But that is the beautiful absurdity of life and the sometimes-at-least result of following a passion to see where it would lead.

While I've been a songwriter, a scribble poet and an occasional blogger, I have never harbored dreams of writing a book. I've learned there are many things in life which are thrust upon us, whether or not we deem ourselves ready. Songs contain a lot of information to digest in three or four minutes, but they are delivered in a package that does not require consuming all of its elements in equal measure and can reveal their deeper nuances on repeated listenings.

Although I am comfortable with the elements of telling a tale to an audience, writing a book is a different animal indeed. As I contemplated the undertaking, the usual naysayers of a creative mind were in full force. The voices in my head that pay no room or board took every opportunity to criticize and cast doubt if given an instant (particularly at 4 a.m.).

What I am though is a "storytender." I consider it my job and my calling to find stories, distill them down to their compelling elements, and like a high-end barkeep, decide what you might most like and pour them out for you to savor.

I am also the grateful heir to a great many tales and many more mysteries, and an eyewitness to a fascinating collection of experiences. Exploring them has brought me into contact with many wonderful people whom I would not have otherwise met. Many of them share a segment of some double helix on this crazy human genome, bequeathed to us by an unknown ancestor, who in turn inherited it from some distant ancestor.

I've certainly seen enormous changes in the music business and the cultural consumption of music since my first album *Traveler* was released in 1995. The digital revolution that started with the compact disc began the demise of album art. What used to hang on the wall of your bedroom was reduced to a four-and-a-quarter-inch square booklet in four-point font size. The advent of digital music and streaming services essentially threw the dirt on the grave. And of course, the changes in music consumption via streaming along with pop culture popularity of worldwide broadcast contests like "The Voice" and "American Idol" have tilted the pendulum back towards singles.

In an industry where the rules are destroyed, rewritten and repeated, I realized that I was under no obligation to do anything specifically in line with some standard practice. As I wrote these songs over the last couple of years, it felt like they fit together to tell a larger story. So even as music consumers are fed an overflowing buffet of singles served up by algorithms, this artist decided to make an album, carefully curated and sequenced, just as they were when wax dinosaurs ruled the music world.

As I planned the recording, I realized containerless delivery of digital music meant that I was no longer constrained to fit everything into the confines of a CD jewel box. But I was also stumped about how to best share the context and experiences that gave rise to the songs, as well as important information about the recording.

The idea of writing a book arose from a session with four trusted friends who are kind enough to serve as a sounding board and offer opinions drawn from their varied experiences. We sat around my kitchen table after dinner with an open bottle or two discussing how I might develop many of these experiences into a keynote presentation. My original thought was a simple pamphlet, mostly presenting the standard liner notes with a little more backstory to each song, in big enough print for eyes of any age. However, many keynote speakers embed their stories, philosophies and secret sauces into books that provide them a platform as well as a tangible take-home for attendees. So instead of a few pages about the backstory to the recording, my friends quickly convinced me to consider developing something similarly robust, which is what you are now reading.

While it can certainly be read as a stand-alone piece, this book and the music album are companions born of the same experiences and essentially created side by side.

For music - the language of my life - to intertwine with the story of my life seems more natural in hindsight than it would have a few years ago. Like millions of other people, my curiosity about my ancestry sparked into flame with the easy availability of online record searches and a magnificent new suite of tools to sift through DNA evidence. Like any good scientist, I got intrigued and followed the clues - literally tens of thousands of clues.

What started as an interest in my musical inheritance evolved into an obsessive hobby and a full-fledged occupation. And frankly, it was far more exciting to track down a stray ancestor than send a dozen emails politely trying to convince concert presenters why they should hire me out of a hundred million singer/songwriters. I also often found it far more satisfying and productive.

My grandmother Madeleine Warner McKnight is truly the main character in this journey. I was her only grandson, and Gram had nearly 90 good years before her memory faded away. She inherited her family history and kept it fastidiously, while also trying to pass it on to her rambunctious and mostly disinterested grandchildren. Gram told us stories whenever she had our attention, and we all grew up with a collection of odd names and the notion that we were somehow related to Robin Hood.

Midway through writing this book, my cousin Claudia emailed me that during a recent move, she had found an old album of Gram's labeled simply "Family History." She was completely unaware that she had it, how she got it or most anything else about it. But she knew I would find it more than a bit useful, as well as incredibly fascinating. In addition to being a treasure trove in its own right, it catalyzed several other miscellaneous findings and remembrances that have added depth and connection to my grandmother's story.

One of the many benefits of having colonial American ancestry is that millions of other people do too, and when I joined Ancestry.com in the autumn of 2012, it seemed most of them already had well-documented family trees and headstone pictures. Late one winter night I was exploring yet another of the branches of my grandmother's family tree, a relaxed journey of curiosity guided by a lovely dram of sipping whiskey. I had followed Ancestry's shaky leaf hints to discover my 10-times great (10G) grandfather, an English soldier and sailor named Nathaniel Turner who became one of the founders of New Haven in the Connecticut Colony. There were narratives about him in some of the biographies of early New England colonists, and I learned that he had met a tragic demise at sea in 1646, a tale preserved in Henry Wadsworth Longfellow's epic poem "The Phantom Ship."

I mulled over this startling discovery for a few minutes before I realized that it was a good thing that he had married and had children before he set foot on that ill-fated vessel, for without him I wouldn't be here.

And how mysterious and magical is that? Sure, we can look back at our family tree and say, "That's how I got here." But what if it didn't happen that way? We each have 4,096 10G grandparents. There may be some duplication of course, particularly in smaller communities where there are often several marriages between families. But every one of those places on your tree is necessary for you to be here.

What if you were to go back in time, and someone offered you a bet - to randomly choose 4,096 people to pair off and procreate, and that 400-500 years later you would be standing here the result of all of their genetic success? The odds must be something like a bazillion to one. Would you take that bet? I'd bet no, but yet, here you are.

The truth is that each and every one of us can be viewed through THAT lens too. That we beat those bazillion to one odds and dodged dozens of bullets, sinking ships and childhood diseases. That through miracles of biology we each inherited those stories regardless of our standing in the world - the powerful, the powerless, the leavers and the left behind. And somehow, miraculously, magically and mysteriously here we are, I the writer and you the reader. When you see your unique existence this way, it tends to affect your outlook on life considerably. By this point, it seemed that maybe I did have something worth sharing in a book.

The final impetus for writing these many pages came seemingly from the Great Beyond. I met my cousin Melody through Ancestry DNA matching, not long after she had suddenly and tragically lost both of her parents in a two-day period. We found that Gram and her grandmother were first cousins who lived in the same hometown and knew each other well. But for the most part, Melody knew very little of her grandmother's family heritage. Despite living more than 1,000 miles apart, we have become close in the way that I have with several of my previously unknown and far-flung cousins. As part of processing the intense grief after her parents' unexpected passings, she had written and published *Transforming Grief Through the Elements.* For her to find us - her family - in the wake of those losses was an additional healing balm for her.

Melody and I met for an afternoon during my Florida tour in January of 2019. We spent our time eagerly sharing family stories and comparing notes like we'd known each other forever. A few weeks later as the idea started to take shape, I told her I was contemplating writing a book, and she enthusiastically offered to publish it. She knew all the ropes that I didn't, how to do all the new things that would be required, and she wanted to help me bring it to life. And suddenly the guy who had no idea how to write a book had a guide to lovingly lead the way.

It was like Gram had sent along one of her family to help me, just when I needed it most.

> ***HOW TO USE THIS BOOK: If you are researching your own family history, I have included a lot of information and resources that you may find helpful. At the end of each chapter is a section entitled Your Treasure, containing some handy hints and suggestions. While it is not intended to be a complete "How to Do Your Family History" compendium, I've collected a lot of valuable tools and techniques I've learned and used from many sources into Appendices C (collecting and saving your stories) and D (using DNA evidence). My hope is that my experiences serve as a lens to see how you might put these tools and ideas to your own good use, should you choose to do so.***

ANDREW McKNIGHT
TREASURES IN MY CHEST

CHAPTER 1
The Spark

What compelled me to spend so many long dark winter nights sleuthing the trail of dead people?

The smoke and haze hung thick in the valley ahead. Ordered into action to guard the advancing left flank, the new recruits marched in a ragged blue line down into the holler, alongside hundreds of worn and weary-faced men. Just six weeks removed from their enthusiastic enlistment in the Union Army and collecting Abe Lincoln's hundred-dollar bounty, the senses of the native sons of Bristol, Connecticut, were now overwhelmed with their first exposure to combat in these cornfields three hundred miles from home. They were far too inexperienced to be here, yet here they were, desperately needed as the survival of a nation hung in the balance. Held in reserve in the woods in case of emergency, since daybreak they had been hearing the echoes of dying men screaming in the distance, punctuated by a near-constant boom and roar of artillery.

Under the mid-afternoon sun, the advancing army moved across the undulating landscape like a blue horde of insects. As they moved through the narrow defile, a rumble grew on their left. Emerging into view from the approaching dust cloud was a large force of men in sharp formation moving swiftly to intercept the blue-clad army, the stars and stripes flying at their head and sunlight glinting on their smooth metal rifle barrels. But beneath the familiar flag marched hundreds of hardened men, clearly bent on a different objective.

The blue-coated soldiers turned to face this unexpected new threat. The menacing forward legions, their gray uniforms now close enough to see, took aim as one under a shouted order. The gleaming rifles erupted in a line of flame as they fired, and then the Confederates charged with a collective, fearsome blood-curdling scream. All over the valley men who a few weeks ago had been merchants, clockmakers, and laborers fell in writhing cries of anguish in front of their onslaught. As the grey horde charged, some of the men of Bristol took cover and tried to rally for a moment, while others, with but a single drill in their three short weeks of training, fled in terror in any direction away from those rebel yells and blazing guns.

I sat bolt upright. Those moments on the field had been so vivid. The park historian today described the role of my great-great-great-grandfather Aretas Culver's unit, comprised of new recruits from his hometown of Bristol and pressed into service in the final stages of the Battle of Antietam. While I had been to Antietam National Battlefield many times, I had never known of my ancestor or his unit's role at a critical stage of the battle. Today our mission had been to walk in their footsteps on the battlefield, a part of the story I had rarely explored.

"What piqued the interest in your family history?" I hear it regularly, and it's a great question. My answer is pretty simple: because it was handed to me. When I reached a point in my life where I started wondering about my family history, the seeds were already in my possession and richly fertilized by the cool technology at my fingertips.

Thanks in large part to growing up with Gram's stories, I knew little bits and pieces about my ancestry and heritage. But today was different. In a flicker of REM sleep, a few moments of my ancestor's story came into sharp and terrifying relief in an emotional way I had never experienced.

It is an age-old human inclination to wonder where we come from and how we got here. Maybe we've inherited heirlooms from elderly relatives, or a grandparent keeps a family tree with a variety of tantalizing mysteries. Many family searchers are adoptees searching for their birth origins. In my case, my childhood was infused with stories and mysteries about this unseen web of dead people that somehow connected to me. It made little or no sense to me then. Most of us who are fortunate enough to reach middle-age eventually turn our thoughts to our past and our legacy, particularly when we take our place as the "bridge" generation between our children and our aging parents.

I believe that understanding our ancestors is vital to understanding ourselves. We all got here through the triumphs and tragedies of our ancestors, and we all have an equal number of positions on our genetic family tree – known or unknown. Born to known lineage, adopted or anonymously parented somehow in passion or rage, biology is biology.

Genealogists are not all that different from people who obsess over dog breeds, collecting specific types of sports cards or knick-knacks, knitting, etc. We collect into communities based on common interests quite readily in the era of social media. And we share our enthusiasm for the latest developments and improvements that further fuel our passion.

Whatever leads you to journey through these pages, welcome to what I define as the first Golden Age of Genealogy. Never before have we had so many powerful tools - from cheap DNA test kits to literally tens of millions of digitized old records around the world, easily searchable on the Internet. You can learn more about your family history from the comfort of your own home than has ever been possible.

I know some people aren't remotely interested in their ancestry. Painful family stories including abandonment or non-consensual origins, or even just not wanting to "rock the boat" in a happy life as an adoptee, will often dampen any enthusiasm for the search. Our deeper genealogies can also bear many scars. Some have lost a generation or more in slaughter and the worst of human behavior, while others have lost many generations under oppressive enslavement. Some people in the past did horrible things, as do people in the present.

And let's face it, people can wind up living really messy lives that adversely affect their loved ones and close relationships. Though we are all born naked, helpless and innocent, it is too easy for the wrecking ball to swing back-and-forth through a chaotic life, careening ever more out of control. This goes for our living relatives as well as our long-deceased forebearers. There are plenty of people in the world who are far from easy to be around, just as there were "back in the day." Without having walked a mile in their lives, I consider most of the circumstances and choices of my ancestors are not for me to judge, but to interpret in the context of their times to the best of my abilities.

Family history isn't for everyone. But family is part of every one of us, bar none. Genetic material has been passed on for millennia upon millennia, mixed up with each generation and blended into something new. Over the course of our lives, we may have no contact with the people who passed their genes onto us. But we still have those genes, and wound within them is the stuff that made their stories possible - and ours. It is good to keep close that we are not responsible for their sins and failings, nor are we entitled to anything other than admiring their triumphs and accomplishments from afar.

The journey of discovery is much like doing a three-dimensional puzzle online with a bunch of strangers. We have our pieces, and thanks to the powerful intersection of DNA evidence and the Internet, family members drop into our lap with new pieces. When we get really lucky, our little family team might assemble a whole corner or side, and bring an element of our ancestry into sharp focus.

I use the word construct **ancestories** a lot. While I wouldn't claim to have coined the term, and have seen it here and there since it first came to mind, it does succinctly bring the why into focus. The stories of our ancestors are why we are here - the good, the bad and the all between. While we would rather not claim the notorious ax murderer who happened to be a four-time great grandfather as kin, that whole nature versus nurture thing is pretty complicated as well.

It is difficult to know how childhood traumas and upheavals shape the lives of living people we know well, let alone the ancestors of whom we get at best a few snapshots and a fleeting glimpse. The context of the time period is essential to interpreting the actions of our ancestors. Their understanding of their world was different, as was the world in which they lived.

Setting biological drive aside, what great cosmic thunderbolt produced that ripple effect some half a millennia ago to somehow cause those 4,096 people who represent your ten-times great (10G) grandparents to pair up and procreate? And through all the possible outcomes of choices and circumstances in every individual's long or tragically-brief life, what were the odds that their DNA would pass down through those 13 generations between to wind up in you, the only person who ever lived with just that specific, magical elixir in their cells?

Whether it is in spite of their genetic contributions or because of them, they are a part of who we are, whether we know it or not. The brilliant, resourceful, one-of-a-kind you would not be here without their contribution. And that is enough for me to appreciate them, rogues and all.

Let me be blunt. The work of discovery is never done and never will be. We will never get to the bottom of this because there is no tangible bottom to discover. We know what we know, we learn what we learn. Through diligence and occasional good luck, we figure out some things that reveal more about where we came from and who we are. But the farther back in time we go, the less certainty we can have about the truth. We have theories, and we have evidence, and sometimes there isn't enough of the latter to even consider the viability of the former. Sooner or later, the trail goes cold. Maybe that's just until technology, record discovery and digitization and subsequent research enable us to make new discoveries even further back. Or perhaps it's just as acceptable to let our imagination take it from there.

You are bound to your ancestors by blood, but also by common experience. Who has not turned up their eyes toward the moon and the stars? Know that your ancestors, long ago and unknown to you, gazed up at the sky in wonder at that very same lunar face looking back at you.

And know too, that given the immense distance that starlight must travel to reach us, that some of the twinkling lights you see in that sky radiated out from those stars during their lifetime, and are just reaching your eyes right now. The same drops of water in the ocean that they crossed have been through countless hydrologic cycles since, as clouds, rain and snow, raging torrent, a mud puddle, and ocean again.

But that's not all. These humans saw sunrises and sunsets. They experienced weather events - blistering hot days and summer thunderstorms, early frosts and raging blizzards. They dealt with drought and scarcity as surely as they were occasionally overflowing with bounty. They must have marveled at nature's beauty, like wildflowers in spring or the turning of leaves in autumn.

They were attracted to mates and fell in love, and likely faced the rejection of the would-be and wished-for lovers, too. They experienced agony and the miracle of childbirth, and the hormone-laden angsts and conflicts of young teens. And they learned the soul-ripping heartbreak of sudden and unexpected loss, as well as the gradual failing and passing of revered elders. They sought to make sense of the unexplainable, unimaginable and unforgivable.

Their daily lives and routines had similarities too, for we did not just recently become creatures of habit, nor is it sheer happenstance. They'd wake in the morning, likely needing to answer an urgent and demanding call of nature. They prepared and ate food, and likely experienced and expressed gratitude for having food to eat. They had children who needed attention, feeding, clothing and of course, parenting. They caught colds, had accidents, broke bones and tore ligaments, and suffered intestinal distresses. While they may not have been literate, they had stories of their existence that had been told to them, and stories of their own that they passed on to their children. They hummed and whistled, sighed, cried, yelled, argued, affirmed and in their own way, had many of the same sorts of conversations and conflicts with their intimates that we do.

All that isn't to say they would recognize us, walking down the street with our gaze locked down onto our smartphone. Myriad things about our time would be incomprehensible and astonishing. They'd recognize the species, and certain traits about us, and perhaps even understand some of our body language, gestures and verbiage. A typical modern city with its cornucopia of skin hues, attire, shapes and sizes would no doubt be sensory overload to our ancestors who never ventured beyond their relatively small geographic area.

So these wonderful strange people of our past, these people who might be nothing more to us than a name, perhaps a birth date, and maybe one small fact we uncovered in a precious written record somewhere - these people were a lot like you and me. Or more to the point, we are a lot like them. We carry them swimming in the cells of our bodies as surely as water, nutrients, vitamins, toxins and all of the other stuff we accumulate in our human journey on this wild planet.

Let us gaze in the opposite direction for a moment. As we generate reams of digital data with the most mundane of our daily actions in the modern world, it is impossible to consider that someday we might be hard to track. I'd wager most people don't give a lot of thought to legacy during their lives, and when they do, it is fairly specifically-focused.

But imagine for a moment, that two or three centuries from now, a descendant from your family is researching you. That person, not yet a gleam in anyone's eye, is tracking a mystery in their family that is an obsession, and you are in some way a key part. You might wonder what could possibly be mysterious about you and your existence, for surely there is a trail that would be easy enough to follow. Maybe so.

Or, maybe not. For just as our ancestors would be shell-shocked by five minutes in our time, so too is our descendants' future largely unimaginable to us. We have no way of knowing what will survive our time, or what they will value in theirs.

Consider that the iPhone was first released in 2007. To say that this smartphone/portable computer technology has changed our lives, our society and perhaps even our very brains isn't hyperbole. We can easily see many of the ramifications as we look back over just the past few years.

A piano-sized planetary probe was launched in 2006 to fly by the planetoid Pluto nine-and-a-half years later. The Pluto of my school science textbooks was a never-seen, dimly understood world of ice described usually by a hollow circle. When New Horizons passed Pluto in July of 2015, it sent back a tiny data stream over 3 billion miles, which was then compiled into richly detailed pictures that anybody could view on those smartphones. If such a future would be nearly impossible to imagine even two or three decades ago, how could we have even an inkling how far past a Star Trek future our descendants will be in two or three centuries?

The point of all this is, of course, impermanence. Surely new developments will leave our time and tools in the dust of obsolescence in the decades to come. Nothing about any ancestor's time or ours was/is constant, and yet, our place in our genetic line is assured from the moment of our birth.

Someday your curious family historian of the future might carefully peruse and parse whatever digital detritus survives two or three hundred years from now, looking eagerly for clues as to what sort of person you might have been. Perhaps their holy grail might be the duck-faced selfie captured on your iPhone, and that the 2010s will be the time when such a valuable type of evidence first became available. They may even wish they had known you. Imagine that!

That sounds like an excellent self-validation mantra to keep handy for any moment you might struggle seeing your own special and unique worth.

> ***YOUR TREASURE: What records do you have of your family history? Now is a great time to start keeping good notes including the sources of what you do know. Store the papers and photos safely, even if you don't have time to go through them in detail. Take extra care to record everything you know about people and dates of your photos on a sticky note or paper taped carefully to the reverse side of the photo.***

CHAPTER 2
The Quest

My grandmother started me on this journey, and her lifelong hometown has been a touchstone throughout.

It's all my grandmother's fault. My father's mother Madeleine Warner was proud of her family heritage and deep roots in her hometown of Bristol, Connecticut. She was a member of the Daughters of the American Revolution (DAR) and dedicated many volunteer hours to help keep the doors open at the Bristol Historical Society. Of course, because of that I knew that some of our ancestors fought in the Revolution. And thus, all of her grandkids were compelled at some point to write out our lineage connecting us to those unruly ancestors.

Gram married my grandfather Andrew McKnight in 1926, on the same date that my daughter Madeleine would be born many decades later. Gram and Gramp seemed an odd couple to me, living essentially separate lives under the same roof going on about their business. They were my main close contact with "old people." So I grew up thinking that all old people just naturally did weird things like sleeping in separate rooms and packing their own shotgun shells. They'd intersect at regular points during the day and the week, and the rest of the time they just kept busy with their varied activities and commitments. In hindsight, they were about as different as two people could be, but they tolerated each other in marriage for over sixty years.

On Thanksgiving and Christmas we would pile into the family car and make the hour-and-a-half trip to Bristol, with me dressed nicely (against my will) in shined shoes and uncomfortable clothes and my hair neatly combed. It might have been in some vain hope that I wouldn't play ball outside and then track mud in my grandparents house, which was inevitable since I was usually bored shortly after arriving. Aided by the adult ladies, Gram would cook the turkey and all the appropriate fixings in her tiny kitchen. It was isolated from the rest of the house, so a kid venturing into that space in search of a pre-dinner snack might well end up with a small chore to help get dinner on the table. While the photographs from those early holiday tables are mostly in black and white, the memories come in a blurry haze of color. It was a tradition that not even miserable late autumn New England weather could stop.

Gram's people probably didn't come on the Mayflower, but they got here in those early colonial New England years. Apparently they waited until everything was all set up and ready before they made the three-month voyage from mother England. Her people were among those in the 1630s who headed inland from the Massachusetts coast, ultimately settling some of the first towns of Connecticut.

I believe I was eight or nine when my turn came to fill out those paper DAR genealogy forms that Gram gave us. I dutifully scratched with my dull #2 pencil and, with marginal legibility, wrote the names, dates and birthplaces of Gram's ancestors - the Culvers, Greenes, Warners and Cranes. It was strange nonsense to a young boy who was much more interested in collecting baseball cards with names like Aaron, Rose and Yazstremski. None of my ancestors ever played in the major leagues, so why would I care about them?

Both of my parents were public school teachers. Each year they survived another ten-month sentence of tolerating classrooms overflowing with lots of kids who didn't want to be there, many facing their own tough circumstances at home and acting out accordingly. So once my parents were paroled for the summer, we were all bumping into each other in our tiny house. While living "out in the sticks" was great for tromping around and getting lost for a couple of hours exploring, it wasn't so great for finding a catching partner, or someone to wing fastballs across a flat stone serving as home plate. And by summer break, my parents' reservoir of patience with a bored and energetic boy demanding attention and activity was rather depleted.

Thankfully, my grandparents would intervene and come collect me for a few days visit to Bristol, perhaps with some vested interest in me surviving to adulthood and someday passing on their genes. I knew two things for sure when I'd visit my grandparents - there were a LOT fewer rules, and they were a lot less rigidly enforced. As soon as I got word that Gram and Gramp were coming to pick me up, my bag was packed ten minutes later, and I'd be sitting out on the front stoop awaiting the arrival of their drab late model sedan.

Bristol is about 70 miles west of my parents' house, in the steep sandstone ridges of central Connecticut. The ride to their house felt like a trip to an alien world, leaving behind the familiar woods and small town church steeples of the east to cross the Connecticut River into a land of rust red rock ledges and bigger cities.

Bristol was strange to me too, a city with wooded neighborhoods and hills dividing it into sections. It was close to Lake Compounce, the amusement park, and the small ski area at Mt. Southington, which my grandfather haunted all winter long. It seemed like my grandparents knew everyone.

My grandparents' house was on a small hill on Central Street in the aptly-named Forestville section, where Gram's family had lived for more generations than any of us could count. Across the street was the library and the Boys Club, where as a child my dad had spent considerable time escaping his parents' house. Right down the street was the center of Forestville and Frank's Stationery, which was my mecca because they sold baseball cards. My grandparents would frequently slide me a couple of quarters AND let me walk down to Frank's by myself.

Of course, while there was certainly some sliver of altruism in my grandparents' willingness to tolerate having a restless grandson around to break the daily routine, Gram had an agenda for me too. There were lots of days when we'd be in the car for a good chunk of the day running errands, often spending time at church, the Historical Society or some other venue where she was volunteering to fill her retirement time with social interaction and helping others. The amount of time I got to hang out with my newly-acquired baseball cards often wasn't as much as I hoped.

Sometimes Gram would host her Bridge Club, or some family or friends, all of whom she wanted to meet her only grandson. There I'd sit for a few minutes, trying to understand exactly what all the fuss was about and why old people always asked such odd questions, had strange hair and smelled funny.

When we were on our own at home or driving out around, Gram was a veritable fountain of stories, often about people with weird names like Aunt Fannie, Cousin Maude, or her friends Betty Manross and Mildred Percival. We would occasionally stop by the Forestville Cemetery to leave flowers on family graves. I found the stones fascinating for ten minutes, and thankfully we usually didn't spend much longer than that. In recent visits, I've come to recognize many of the names on the stones and where they fit into my family. It seems as though much of Gram's world is buried all around her.

Gram was particularly attentive to the stones of her family veterans buried around town, including a Revolutionary War ancestor whose name and final resting place now escapes me. It was cool, but the engraving on the stones was old and hard to read even then. I did think it was pretty neat that we'd had ancestors who'd fought for America's independence, but it didn't seem like they'd been at any of the places I knew – Lexington and Concord, Saratoga or Yorktown.

Despite my childhood inability to understand much of my grandmother's world, we were extremely close well into my late 20s, before her memory failed her and she entered the nursing home. I stashed away the scraps of family lineage she had made me write out and a few family photos. Occasionally I would have a flash of remembrance of some story she had told me. But gradually I grew to regret not having written more down, asking more questions and especially never recording her telling those stories.

At some point in junior high or high school I had one of those assignments where you get one of your elderly relatives to tell you about their life. So Gram scratched out a bit of an autobiography in six pages or so, about her days as a young nurse caring for tuberculosis patients in the 1920s, and teaching young nurses as her career got established. It is in her own hand, with her own humble humor reflecting on her life, referencing herself at age 87 as being as useful as a third left hind wheel. Gram was proud of her life's work, but she sure as heck wasn't going to toot her own horn about it.

I keep the tangible treasures of my family, including Gram's handwritten story, in a small wooden chest that she handed down to me. It is a seaman's chest, owned by one of her ancestors, whose name likely will remain forever unknown to the living. I've never been able to identify a plausible candidate in her family who lived in the last 150 years, but it could possibly predate that.

Needless to say, those six pages are an irreplaceable jewel indeed, her life in her own words. I've scanned them and saved them in at least a dozen places. I knew even then I valued the gift of those pages because she had written them for me. But now in my middle age, they speak to me in a deeper way that I can't describe. The knowledge that she wrote them, carefully looked them over and made an edit here or there, keeps her close.

No one in my world would have appreciated my journey these past few years more than Gram. To know she had planted a seed of interest in family stories surely would have given her deep satisfaction. I often feel her presence as a silent partner in the research, especially when working late into a winter evening in the time of year that I've come to appreciate as the "long nights of genealogy season." I know I have found countless stories in her family that she could have told me herself, and a few I'm fairly certain would have been completely new to her.

When I was eight, Gram crotcheted an afghan for me from my favorite colors, and of soft wool that I favored over the more durable scratchy wools she typically used. Of course, I still have it and I do try to take good care of it. I often joke that it is magic wool, for rarely has anything warmed my cold body faster. And since we keep the thermostat low overnight during the winter, I often curl up under that afghan to sleep. I suppose Gram really does keep me company quite often.

Gram was certainly a social creature. She loved meeting people and keeping a wide circle of friends. She had the gift of putting people at ease with kindness and humor. Gram also kept close relations with many of my grandfather's family, frequently to his chagrin as he did his best to avoid most of them. I believe she would have enjoyed most of all my discovery of and connection to living relatives who share some piece of her ancestry.

And in that regard as much as anything bestowed on me from her DNA, I am indeed her kin.

> ***YOUR TREASURE: A basic family tree is an essential tool in learning about your ancestors, but often we start without knowing much of anything at all – more on that in later chapters. Whatever you DO know, take a few minutes to organize it into a family tree sheet that you can refer to as you learn new things. I recommend that you write in pencil or an online sheet, for you will often discover that there is some uncertainty in even the most basic life and death records as well as spellings. A tree that includes even your eight great-grandparents is a tremendous start! Using free online resources like FamilySearch.org can help you build back in time with helpful hints, but take the time to be 100% sure that hint is for your John Smith.***

CHAPTER 3
Plainfield

How the place of my childhood shaped my understanding of leaving "home" behind.

If you've never been to the northeastern part of the Connecticut, you would likely struggle to reconcile the dichotomy between the rural hill country along the eastern border and the Wall Street suburbs of the Gold Coast (experienced by most from the never-ending parking lot known as Interstate 95). The "Quiet Corner" is not a place one goes by accident, or by just passing through. It takes work to get there. And for some of us, it would have been too much work to stay.

The Quinebaug River drains south from central Massachusetts through glacier-sculpted valleys towards Long Island Sound. The towns along the river boomed and collapsed with the mills of the early 20th century, leaving in place the French Canadian immigrants who came to operate their massive machinery. In the hills, people farmed and many were still largely self-sufficient. My dad took a job teaching high school science in a town that valued neither school nor science, particularly when it came time to vote on annual school budget proposals.

I spent my childhood escaping reality as best I could in those hills. We lived on a high ridge east of the town of Plainfield, tethered to it by two miles of a winding road that continued through ever less populated hills and into the swamp forests of western Rhode Island. What a century previous had been cleared farm fields bordered by walls of rounded stones was now reclaimed by New England's resurgent forests.

We had no known family connections to the hills of the borderlands, but there were plenty of intriguing stories hidden there. Our backyard practically touched the northern border of Pachaug State Forest, Connecticut's largest. One of the attributes of spending your first decade as an only child is a fertile and occasionally overactive imagination. With only a couple of neighbors, I was pretty much responsible for my own entertainment. Those woods provided plenty.

Traipsing through the forests and fields (and trespassing on every absentee landowner's unused property), the echoes of the past were occasionally visible underfoot - most typically an abandoned stone foundation and perhaps some other evidence of prior habitation. The family of a childhood friend had been there probably since the community sprang up around 1700, and just down the hill was the overgrown cemetery marking the final resting place for many of his ancestors.

Life was quite different down in town, one of the many mill towns that sprang up along the Quinebaug. Probably 90% of my schoolmates descended from the Quebecois, who migrated south in the late 1800s, desperate for work and a better life. They found at least the former in the southern New England textile mills. They lived in the duplex houses built by the mill owners.

Many of my classmates - their descendants - grew up in those houses. Most of my peers went to St. John's or All Hallow's Catholic churches, observed Lent, disliked catechism, had *memeres* and *peperes* who still spoke the language of their ancestors, and had multiple sets of cousins from several other large families in town. While some went away to college, it was far more common to tolerate school for a while and then get a job doing manual labor.

The mills of my childhood were hulking shells of their heyday, employing at best a couple hundred people hanging on to what work remained in the death throes of the northern textile industry, while worrying what the hell their kids were going to do when they grew up.

For many of them upon coming of age, the military was one of the few viable options.

Plainfield was also, in a word, white. While nearby towns saw an influx of Vietnamese, Cambodians and Laotians escaping the war in southeast Asia, for whatever reason they bypassed Plainfield. Through most of my childhood there were a couple of families from Puerto Rico and one African-American family living somewhere within the town's four villages. Those families, plus the Jewish family that owned the local five and dime, and a Korean family I got to know for the couple of years they lived down the road from me, were pretty much the sum of my exposure to ethnic and religious diversity.

One day my buddy Geno and I skipped out of high school and were driving around and goofing off. When we pulled up to the traffic light to turn down his road, at the corner was a guy in a white robe and hood passing out leaflets. Apparently at least one local Klansman thought that the working white folks of eastern Connecticut would have sympathy enough for their ideology to be out in broad daylight.

Of course, I was not from a big Catholic family, which made fitting in a challenge at times. My seemingly small family lived in busy towns and small cities in central Connecticut. We would occasionally get visits from aunts, uncles and cousins when my parents would host a backyard barbecue. The relations generally got out of the car stiff and sore, and a bit distressed that their interminable journey east brought them to the deep wilderness at the edge of the earth. They had lots of malls with big stores and shopping, and good high school sports teams. Their lives were about as different from mine as anything I could imagine.

We were among the handful of families who attended the big white Congregational church with the steeple. Dad never went to church, so this was my mother's province and ritual. Every Sunday morning we sat in an unforgiving wooden pew in the presence of the Reverend Gordon Johnson's sonorous voice and gentle manner. I fought sleep as best as I could. In Sunday School we didn't talk much about why we were different from all the Catholic families, but occasional news on TV of IRA violence between Catholics and Protestants in Northern Ireland helped infuse our perceptions of each other with a bit of mistrust. Living in a predominantly Catholic town meant I attended a lot of weddings, funerals and the occasional Christmas Eve mass in those churches. On several occasions I heard directly from the priest's lips at the altar microphone a litany of reasons why non-Catholics were somehow lesser in the eyes of God. It wasn't exactly encouragement to visit again.

So while I grew up outside of Plainfield, I truly LIVED outside of Plainfield without much to keep me there when the opportunity came to leave. Of course my parents stayed, because that's what their generation did. You bought or built your house, and you held onto it as long as your faculties and circumstances allowed. Generational wanderlust was mostly the calling of the young, whether to escape their circumstances or their families, or to seek an adventure somewhere away. It was enticing to escape the mills, or whatever and wherever hard life had enveloped their parents.

For a kid with a latent interest in history, eastern Connecticut presented another problem - nothing important seemed to have happened there. I knew a very little bit of the massacre of the Pequots during the Pequot War (1636-8), but that was 60 years before Plainfield was even settled. All the glamour moments of the American Revolution (so considered because they made it into my grade school history books) happened in other states. Plainfield's lone claim to any fame of interest to me was the march of the French Army under Le Comte de Rochambeau in late June 1781, on their way to New York from Newport, Rhode Island, to link up with George Washington and the Continental Army en route to the siege of Yorktown. I dimly remember a few folks dressed in French uniforms recreating the march sometime around the American Bicentennial year (1976). George Washington wasn't there either time, so I didn't care.

Of course back then, none of us knew anything of what we have recently learned - that in 1699 my father's 6X great grandfather Jonathan Crane traveled some 20 miles east across tribal lands from the settlement of Windham to inspect and approve the site for the original meetinghouse in the new settlement of Plainfield - or that another of his "six greats" married and lived out her life just across the river in Canterbury in the early 1700s. We didn't know then that a large number of colonial settlers in the area where the Thames River rises from the Quinebaug, Shetucket and Yantic rivers, were my mother's ancestors - or that both sides of my family had raised big families in central Connecticut for centuries, my mom had a ton of cousins, and my dad's mother was related to half the city of Bristol.

Anything that you were certain of as a kid tramping about in the woods turned out to be a whole lot less black and white, and a whole lot more of a spectrum of gray.

> ***YOUR TREASURE: The best time to plan your basic organization is when you get started, but the second best time is now. Start simple and stick to it – make a folder for each of your four grandparents families and keep papers and notes in there as you go. Keep whatever work you do on the computer in a similar hierarchy to make it easier when you are searching for***

something. An online note program like Apple Notes or Evernote is a great resource for collecting details about your closer ancestors. Use a spreadsheet or table to make a simple timeline (you'll eventually have several timelines for different purposes). Most of all, backup your work to a secure Internet location in case disaster strikes your device or worse, your house!

CHAPTER 4
Music in My Family

Drilling down into the "inheritance" of my life language.

I don't know a life that doesn't include music in a big way. My parents say that as a toddler I'd stand in front of Dad's piano plinking away and that I'd find the harmony notes and play them together. Looking back on it, they might have been a bit glassy-eyed about their first-born, since it's pretty hard NOT to find some harmony when hammering two different notes in the white keys of C. But I also wouldn't deny that I grew up absorbing the language of music aurally and orally.

Teaching in a fairly poor town in rural eastern Connecticut was a long distance away from lucrative. To have anything more than the basics required some extra income, so Dad took the notion of "moonlighting" to heart. My earliest memories include him rehearsing with his band one night a week in our basement. I'd hear them picking apart the popular rock and country tunes of the early 1970s, working on vocal harmonies, intros and outros.

Before I was born, my dad played guitar. About the time I came along, he switched to the piano because there were lots of good guitarists in the area. He figured keyboard players were more likely to get opportunities in local bands, and like he was about a great many things, he was correct.

When I was eight, we traveled down to Orlando for a couple of weeks so Dad's band could do a run of shows backing Brenda Byers, a country singer who split her time between eastern Connecticut and the Florida nightclub circuit. We stayed in a motel most of the time, and had our days free to do family stuff, while dad went off to his show at Monte's in Winter Park at night. It was how I got to Disney World.

My grandmother's only sibling, her younger sister Marjorie, was a classical pianist. She would come to visit when I was a kid and I loved to hear her play. It was so different than anything dad would play on the piano. She was kind and and adoring of kids, and I so loved her visits, infrequent though they were.

Aunt Marj contracted leukemia when I was a tween, and I never got to see her or to say goodbye. It is a lingering regret. I especially wish that I could have talked with her about music as I developed my own voice.

My dad is a far different musician than I, and my sister Aly is far different from both of us. It is a bit amazing how three people who lived under the same roof for so long could have developed such wildly different musical abilities and language. Dad never learned to read music, so he developed his ear and his unique way of charting out songs by placing the chord changes over the lyrics. He learned to improvise fairly naturally, while Aly learned piano and flute via more standard lessons and sheet music. She never learned to improvise comfortably, but she can somehow transpose a flute melody she's reading into another key "on the fly."

Then there's me. I often say it was a little like growing up in an indigenous culture with no written language. I learned to recognize sounds and pitches, even though I couldn't play them smoothly. I heard how the grownups did it in my basement every week, and I knew that someday I wanted into that club.

By the time I started dabbling, Dad's old Sears Silvertone guitar and classical guitar were my two options. The strings on each were about an inch off the fretboard at that point, and they were nearly impossible for my small fingers to make sound good. My early attempts were exercises in discomfort and frustration.

But I saw Dad's guitar player Peter at rehearsal each week coaxing magic out of his beautiful Fender Stratocaster electric guitar and what we'd now call a vintage amp.

By the time I was about 12, I developed a higher tolerance to finger pain and discomfort, but I still struggled to play three chords in rhythm with a song. In hindsight, I'd say being self-taught and fully immersed in struggling have probably made me a better teacher.

I also immediately started "making stuff up." When I finally got so I could hold an E chord down, I started playing around and moving it up and down the neck. It was easier than changing to another chord shape, so I just strummed along with the chiming open strings looking for things that sounded good. I dutifully made notes of what I did so everyone would know this was the song I wrote, and this was how it went.

I coveted an electric guitar - a Fender Stratocaster. I had no real hope of coming up with the hundreds of dollars it would take to get one, but my parents made it clear that one wasn't going to appear out of thin air. They encouraged me to take a farm job weeding strawberry fields for minimum wage, about $3 an hour. I spent after-school time and weekends on hands and knees in the hot sun, carefully saving my earnings and scraping together what I could. All the while I kept an eye on the "for sale" section of the newspaper classified ads.

One day an ad appeared for a three-year-old Strat, in mint condition from a private seller in the nearby city of Norwich for $300. Dad was at this point willing to add a little money to the pile I had carefully saved. He drove us down to the guy's house to take a look. Polished shiny and new looking, it felt silky sweet as my trembling hands moved excitedly around the neck - especially compared to the impossibly high-stringed guitars I'd been trying to master. We threw our combined $268 and a couple handfuls of chump change on the table in hopes it would do the trick.

It did. I was now the proud owner of a Fender Strat - the real deal, the tool of legendary guitarists from many walks of life. Dad had plenty of amps in the basement, so I could at least play it when I got home. We walked into the basement music room, and I laid the case on the couch and opened it up.

A boy of that age doesn't know much about true love. But looking back, that's exactly what it was. The smooth curves of the body, the three thin pickups and the narrow neck with all the tuning pegs on the uphill side. I was smitten - hard, truly and completely. Practicing was never again going to be an issue. Getting me to stop was the new problem.

If you wanted to play music in eastern Connecticut, you pretty much joined a band or started your own. There wasn't much jazz or classical music to be had, and original music of any sort was even harder to come by. Bars and clubs were where generally the same people drank themselves senseless every weekend, hollering out the same requests, and showcasing their own style of the "drunken shimmy" on the dance floor, which evolved as the night wore on. You played rock, "Top 40" or country music mostly, and the most popular bands played all three. So learning the tunes to sound "just like the record" was a weird badge of honor.

I was kind of thrown into the water, musically speaking. I wound up in a band of adults who needed someone to stand there and strum the guitar, play a few simple riffs here and there, and not stick out. The drinking age was 18 then, and by age 15 I was playing in a lot of those clubs, mixed up with the occasional wedding or private party, where people got dressed up and did essentially the same thing.

My parents always stressed to me that I should plan to choose a different career path because music was often a difficult life and fraught with pitfalls, but that it was a great and fun way to make money on the side. They were right on both counts. A career in performing, recording or composition required an essential skill that I never developed, though I tried mightily - the ability to read and write music. As a sophomore in high school, "playing by ear" regularly earned me a couple hundred bucks on a typical busy weekend playing those bars and weddings.

And of course, I wasn't ready to give that up easily when it came time to go to college. At first I decided to be an English major because I loved writing. But after two weeks I had had enough of that. I made a beeline for the opposite corner of the campus and decided to major in whatever was in that building. It was the School of Engineering.

My path through higher education was a bit stop-and-go for the first few years as I simultaneously chased rock band stardom with my own original band Nor'easter, played in a cover band and worked a job or two pumping gas to make money, AND tried to keep up with engineering classes.

That Strat was my constant companion throughout. I finally settled on majoring in Chemistry and finished my last two undergraduate years at Connecticut College, then went on to the University of Massachusetts for three years to get my Master's in Environmental Engineering. I put myself through the last four years of school playing mostly jam band/classic rock/take-it-out-for-a-while- and-come-on-back bluesy rock with two longtime friends and high school classmates. We deliberately eschewed the "just like the record" approach in favor of arrangements and sounds that more uniquely fit our preferred format of guitar/bass/ drums with everyone singing.

When I finished grad school, I abandoned for good the late nights in smoky bars with their mix of twirling and stumbling drunks, and I bid Plainfield a final adieu. The Strat came to Virginia with me. We're both still here. It's been a part of all of my recordings over the last 25 years. The environmental engineering career was lucrative by comparison, but I said that farewell a long time ago, too.

It is a common question for musicians, "Where did you get your musical ability?" I suppose the short answer is practice and repetition of muscle memory, plus opportunity and exposure. I've known plenty of good musicians who were the outliers in their family, and plenty of good musicians with a family thread too, so I'd probably consider genetics a minor influence at best. Yet people ask it all the time, so there is a perception that one must inherit the ability from one side or the other.

Up until a few years ago I'd attributed most of it to my dad's mother's family because of Aunt Marj, the pianist. After all, I was exposed to her musicality now and then, and she was the only older member of our family on either side who played well. I remembered hearing that my great-grandfather was a fiddler from Scotland, and that my grandfather's estranged sister Margaret was an opera singer, but the only music I heard on Gramp's side of the family was my great Aunt Isabel droning away on her organ. That didn't give me much enthusiasm for their family contribution to my musical inheritance.

Even with all of those possibilities, I don't know if there is such a thing as a "music gene." My own experience might suggest it's just as plausible that simply being immersed in the language of music makes it accessible for the duration of one's life. But between those family stories and the discoveries that were awaiting me, I was about to look into it in detail.

One of the many reasons for writing this book was finding out just how complex that answer might be.

> ***YOUR TREASURE: What talent or creativity seems to have been passed down in your family, or what talent do you have? Have you wondered if it might have been an inheritance as well as an inclination? When we are lucky enough to have some knowledge of a grandparent or great-grandparent's skills and inclinations, it helps give us context about their lives - and perhaps our own.***

CHAPTER 5
We All Came From Someplace Else

Most all of us have ancestors who traveled great distances for some reason and began new lives. Thanks to the Internet and DNA matching, it has never been easier to connect with the descendants of those they left behind.

It is pretty well accepted that the peopling of North America is a relatively recent phenomenon. While it is generally thought that our ancient human ancestors walked out of Africa into the Middle East and beyond some 70,000 years ago, humans have probably been in North America for some 15-20,000 years. We'd certainly grant native status to those people whose ancestry derives from these aboriginal adventurers. But for the rest of us, our ancestors migrated here much more recently. To be blunt, we are all descendants of immigrants.

In our hyper-politicized America today, the whole concept of immigration is so grotesquely skewed as to barely make sense to an outsider. The truth is that up until very recently, immigration meant there were those who left and those who were left behind. The reasons for these separations ran the gamut from adventurism to enslavement. But at the personal level, these were permanent family separations, not unlike death. Until the late 1800s, it would have been difficult to get any word back to one's family under the best of circumstances. And for those coming in poverty or the chains of enslavement, it was nigh well impossible.

No matter where your origin stories may lead, every ancestral location presents a set of challenges including language, survival of records (assuming they were kept at all) and disruptions from disasters or wars. Countries appear and disappear over the centuries as they are absorbed into empires or achieve some measure of independence. It is relatively safe to say that outside of colonial New England records, Americans are likely to face stiff challenges to learning their ancestry before the 1700s - particularly of their immigrant ancestors' home regions. My mother's roots in Ireland are no exception.

As anyone with Irish heritage and a "wee bit" of family research experience will tell you, there seems to be as many brick walls and mysteries to be found on the Emerald Isle as there are actual records. Then there is the small matter of the sectarian discontent, which has bubbled and boiled there for centuries, most recently in "the Troubles" between the Protestant majority in the northeast and the majority Catholic Republic of Ireland.

To be fair, there's plenty of bad-blooded history to go around. A main challenge of Irish genealogy is there is much to consider when seeking your ancestors. It is often said of the Irish that they left misery for someplace far away, only to sing of how miserable with homesickness they are for the misery they knew. While the Famine or the Great Hunger of 1845-52 that depleted the Irish population is the best known, there were other Irish mass migrations before and after.

At its closest, the northeast coast of Ireland at County Down is but 13 miles across the Irish Channel from southwestern Scotland. Even back around the year 1600, the boat crossing took only three hours. It was made frequently enough that no one bothered to keep travel records, as it was all essentially part of the British Empire. Upon the ascension of King James to the English throne in 1605, Counties Down and Antrim soon began receiving large numbers of Scots, lured by the promise of land and the comfort of a body of water between them and London. Despite his Scottish origin, King

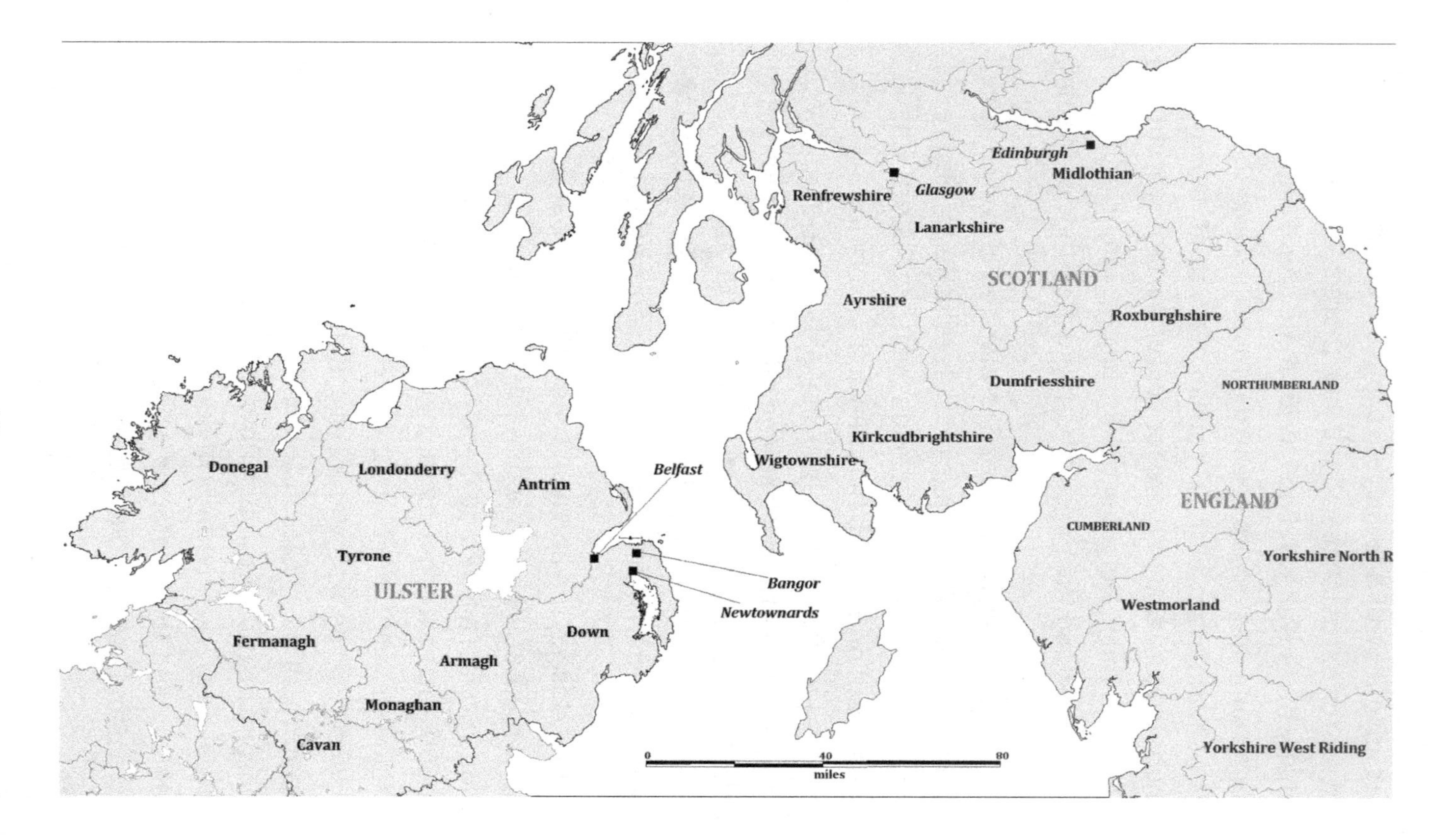

Edinburgh
Midlothian
Renfrewshire
Glasgow
Lanarkshire
SCOTLAND
Ayrshire
Roxburghshire
Dumfriesshire
NORTHUMBERLAND
Kirkcudbrightshire
Wigtownshire
Donegal
Londonderry
Antrim
Belfast
ENGLAND
CUMBERLAND
Tyrone
Yorkshire North R
ULSTER
Bangor
Newtownards
Westmorland
Fermanagh
Down
Armagh
Monaghan
Cavan
0
40
80
miles
Yorkshire West Riding

James liked the idea of having that body of water between him and the more troublesome among his kinsman. And so began the Plantation of Ulster, where the English and Scottish lords who were awarded land in northern Ireland gladly accepted the ornery and industrious Scottish transplants to work it for them, while displacing the native Irish in the process. It was but one of many English insults to Ireland keenly remembered into the 21st century.

The northern nine counties of Ireland are the ancient province of Ulster. Six of those, including Down and Antrim, are known as Northern Ireland and remain in the United Kingdom. The Scots who grew disillusioned at the lack of opportunity in Ulster started emigrating to North America in large numbers around 1700. Those who came ashore and continued west until they found places in the mountains, to be self-sufficient and unmolested by distant politicians and kings, are those we know as the Scots-Irish. Those that remained in Ireland are often referred to as Ulster Scots.

Up through 1921, the entire island was known as Ireland, but was part of the British Empire. And the Irish spent a great deal of effort over the centuries trying to expel the British, who in turn spent a great deal of effort trying to impose their will (and the Anglican Church) on the Irish.

Of the many great tragedies of Irish sectarian strife, one that most haunts genealogists was the destruction of the Public Records Office in Dublin in 1922, near the end of the revolt that finally established a free and independent Republic of Ireland. Apart from a few fragments, many primary record collections were lost including nearly all the Irish Censuses between 1821 and 1851, and a wealth of local and regional legal documents, records and Anglican Church of Ireland registers.

While the loss of these normally low-hanging fruits typically collected by family researchers is rightly lamented, there are many other types of Irish records readily available and several ongoing digitization projects will eventually bring those records online. Civil record keeping of births, marriages and deaths (BMD) began in 1864. Various Catholic and Presbyterian parish records are also available, at least in person. And because the English were particularly interested in collecting taxes of all sorts, the tax rolls in Griffith's Valuations beginning in 1847 often can offer additional clues about your ancestors.

Of course, if your ancestor is a John Ryan or a Michael Murray, you may be stymied trying to determine which of the 27 identical names on the page is your guy. One must dig deep, take time AND have a little "luck of the Irish" to piece together hints with a paper trail of any sort to learn anything of one's ancestors before the 1860s.

In mid-19th century Ireland, the "American wake" was an all-too familiar ritual in every family - a big gathering to say a final farewell and *bon voyage*. It was a lot like a death, for it was unlikely those left behind would ever know if those who left ever reached land again, let alone somehow managed to survive and make a new life. Ireland suffered a population drain of epic magnitude - one of every two people born in 19th century Ireland lived out their life someplace else.

And my family was no exception. My mother's great-great-grandparents from Counties Cork and Carlow came here in the 1850s, just after the Great Hunger. There is no evidence they ever again had contact with their families back home. At best, we knew where our people lived before they left. It's mostly anecdotal evidence written down and carefully kept by American descendants who knew nothing of the homeland, only that "our people lived there."

By the time we had invented and developed the means to readily communicate across the ocean, we'd lost the knowledge of our family roots. And no amount of trans-Atlantic cable or satellites could replace that lost knowledge. If you knew where to look and how to do so, you might get lucky. Otherwise, it was just family legend.

I didn't grow up holding any known Irish traditions - we weren't Catholic, and let's politely say that my parents aren't culinarily adventurous. In college I started feeling drawn to Irish heritage mostly through the music. I spent an unforgettable St. Patrick's night in Boston with my grad school housemate, hitting a parade of pubs with fiddles and accordions after a delicious corned beef and cabbage dinner in the city's Irish heart of "Southie." It was the first time the Irish rock band U2 played

Boston on St. Patrick's as well, so the city was electric with Celtic energy. It was an epic experience, and so too was the hangover. But the lifelong affinity for my Irish ancestry was sealed then and there.

When I started researching my mother's family history, I realized that her mother's entire ancestry was Irish. And it turned out that there was quite a bit on her father's side too. There we were - our Ryans, Kavanaghs, Murrays and Hennessys. I felt like a man who won the genetic lottery, until I realized that pretty much everyone on Ancestry.com with Irish roots had a tree full of sticks labeled Ryan, Kavanagh, Murray and Hennessy. By the thousands. Most of the sticks vanished into an emerald fog in the mid-1800s.

DNA testing was like splitting a genetic atom - the entire world changed. The first time a new match turned up in Mom's family, and shared DNA with most of our cousins with Ryan ancestry, there were no Americans in their tree. I began to realize these might be relatives who STILL LIVED in the places from where our ancestors departed. With jittery hands, I sent emails that began with, "You share DNA with my mom and several of her cousins, and our family legend has said that our great-great-grandparents came from Sherwood in County Carlow..."

And they answered! "Yes, that's just 20 km from here, and my mum's grandparents had a farm very near there." It was then that it hit me, full-on, straight in the heart, that the descendants of those who left were reconnecting with those left behind for the first time in probably 170 years, directly and in nearly real-time across 3,000 miles of angry Atlantic Ocean.

While I've still not met her in person, my cousin Cathy Callan is a painter living in Athy, Ireland. We've grown close across the distance, sharing in our love of the beauty of the natural world and the inherent common threads of being creative professionals. We've shared our artwork with each other.

Since I first set eyes on the springtime floodplains of Virginia carpeted in pinkish-purple bluebells, I have been in love. The flowers grow atop stalks a foot or so high in great profusion along the flatter stretches of creeks and the nearby Shenandoah River. When the twilight is just right, they create a bluish-pink mist on the riverbank that is simply otherworldly. Along with the blooming redbud and dogwood trees, they make April a true heaven on earth here.

One morning Cathy posted her latest painting on Facebook - a clump of violet purple Irish bluebells. I had no idea they grew in Ireland until that moment, and suddenly my cousin's talent was filling my screen with their gorgeous violet hues. It was one of those transcendent moments of familiar connection with something completely new - one of a great many that I have experienced since I started digging around in my treasure chest.

Thanks to high-resolution digital photography, the Internet, and online ordering of canvas print wall hangings at the local big-box superstore, Cathy's Irish bluebells adorns our living room. A cherished family jewel, her painting faces the window looking out over the small patch of native Virginia bluebells I have cultivated in a wet spot in our yard.

When people ask me why I do this, I think of the painting on my wall, and my cousin far across the sea, and of reconnecting us American immigrants with our families back home after a century-and-a-half. In some way perhaps we are woven back together after all, like those red, blue and white yarns in Gram's afghan.

Your story may be similar, or it may be drastically different. You may never know many of your stories, or you may have been bequeathed a literal treasure chest too overflowing to even process. All their failings and their triumphs were fired by the genetic fuel that made them who they are. Whether you know them or not, they are yours - stories wound into the base pairs coiled in endless double helices in millions of cells upon millions of cells.

DNA is much more than a molecule - it is magical, and you are proof. It's the blueprint for your resilience, your curiosity, the color of your eyes and your genetic Achilles' heel. You shed tens of thousands of cells every day, replaced by tens of thousands more. They all reproduced according to that same one-in-a-bazillion blueprint that is you. It was somehow magically arranged through each generation of recombination and rearrangement, and courtship and hardship, and then passed on, reprocessed and ultimately bequeathed to you.

In my humble opinion, that makes you a miracle. For you are the only one of you ever. Even if you have an identical twin who's remarkably similar in nearly every way, you have your own brain, eyes and ears to process your experiences, in your own uniquely you kind of way.

So put that one in your pocket too. You are a gift - regardless of your people came from, however they got there, whatever they brought with them, inside and out.

> ***YOUR TREASURE: What do you know of your immigrant ancestors? Do you know anything specific about the area or people they left behind? There are many possible resources like civil records (town/parish, county, state/province, etc.) as well as religious, military, land use and other records. What larger events like an economic boom or collapse, war, mass migration or epidemic might have influenced their lives? Learn about the history of a place during the time your ancestors lived there, and you may get insight into how those events might have affected them. Here's another great job for a timeline too.***

CHAPTER 6
DNA is Weird and So Are We

The magic molecule that makes life possible, research interesting and the mind wonder.

DNA (deoxyribonucleic acid) is the hereditary material of most life. Those three letters may have changed the world as much as ABC. First observed and isolated in the mid-1800s, DNA is most commonly connected with Francis Crick and James Watson, who described its weirdly fascinating double helix structure in the early 1950s. Much like the advent of written language, DNA has become a foundational element in our society and culture in myriad ways including medical treatment, law enforcement and of course, genetic genealogy.

I managed to get through basic high school and college biology with a mix of mild interest and resignation. Even though I wound up being a chemistry major, I had at best a cursory understanding of cell biology and this weird twisty DNA molecule that reminded me of some trippy ladder out of *Alice's Adventures in Wonderland*.

Every basic biology course covers the principles of genetic inheritance first described by Augustinian monk Gregor Mendel's study of pea plants in the mid-1800s, as well as later applications to *drosophila melanogaster*, better known as the common fruit fly. I don't remember much of my high school biology class besides the human reproduction part, but I have remembered drosophila to this day, likely due to its prolific kin taking up residence in my kitchen each peach and tomato season. I always thought it would be a cool name for a rock band.

Frankly, DNA is just plain mysterious to me. How one explains the evolution of a crazy double helix molecule from some primordial ooze is beyond my imagining, although I'm sure there are plausible theories beyond my understanding. I wonder about some unseen cosmic hand of divinity many epochs ago, a random lightning strike in a dark pool of just the right stuff, or perhaps a final parting breath of an asteroid impact from a galaxy far, far away.

The mystery gets thicker when you learn that on a sequence-by-sequence basis, more than 99% of the DNA between humans is identical. And to get weirder, we share around 98% of our DNA with a chimpanzee, and perhaps 60% with a banana.

To top it all off, this miracle molecule can self-replicate. To my mind, this is the place where the line between chemistry and biology blends into voodoo.

Your 23 pairs of chromosomes include 22 pairs of *autosomes* and a pair of gender or sex (XY) chromosomes. Chromosomes are made up of genes, which in turn are made up of DNA. Genes are the molecular basis of heredity - the traits we inherit from our parents, and they from theirs, etc. All your genes collectively make up your *genome* - the complete set of genetic instructions found in nearly all your cells, containing all the information needed to grow and maintain you.

Humans have approximately 20,000 genes, and the differences between individuals come from slight variations in these genes. Even though about 99.9% of our autosomal DNA is identical with one another, this still leaves some 3 million potential differences between your genome and anyone else's. Looking around a room full of people, one may conclude those differences are a BIG one-tenth of one percent.

Genes are why we have Uncle Herb's ears, because we both got them from his grandfather Hugh. And they are why mom's hair is curly like her sister Betty's, and not straight like her sister Harriet's. The visible stuff of inheritance is plain to see, and often fascinating (except when we've inherited a less than desirable family feature).

The foundation of genetic genealogy - the basis for the popular common at-home "spit 'n send" DNA test - is that variations in our DNA could be very useful beyond who would inherit blue eyes, dark skin or genetic diseases. These variations we inherited from our parents, and they in turn from theirs, allow us to connect the dots back into our genetic past. We inherit autosomal DNA and a sex chromosome from both our parents equally. But there is a catch.

Most adult cells contain pairs of chromosomes, but sperm and egg cells have only one chromosome per pair. That chromosome has been reshuffled and rearranged somewhat randomly in a process called recombination before being passed on. This makes each sperm or egg cell unique and ensures that each individual has their own unique genetic code. Unless you are an identical twin, there truly is no one else on this bustling, busy, bursting with life planet who is exactly like you. (And even identical twins won't stay truly identical for long.)

Each of your chromosomes is a pair, one from each parent. When your body creates its progeny cells, it will make a new single chromosome during recombination. The resulting sperm or egg cell will usually have 50% of each parent's chromosomes, which had been similarly created from their father and mother.

It's like your parents get to contribute 50% of their DNA to each child, but have no say in WHICH chromosome segments inherited from their own parents are included in that cocktail, or how they will be arranged. While my child has 50% of my DNA, her test shows that she inherited slightly more of it from my mom than my dad. Except for identical twins, each sibling will get a different 50% from each parent than any other siblings. And while having more children increases the chances of passing on more of each grandparent's DNA, it is still likely that many of their DNA segments will not be passed on at all.

While genetic genealogy is incredibly complex and still evolving, any family historian can learn a lot equipped with a very basic level of knowledge. Testing your autosomal DNA with the popular at-home kits is simple, requiring only that you submit a saliva sample or cheek swab via the prepackaged container. When the results arrive a few weeks later, they will include a long list of DNA matches - people who share DNA segments with you and the amount that they match you, a quantity generally expressed in centiMorgans (cM).

These matches are the genealogy goldmine - other users in the company's DNA database who match you on at least one statistically-significant segment on a single chromosome. Unless you are related to the person through both of your parents, you will have inherited that segment through one parent or the other. Close relatives will match you strongly, while more distant relatives match at lower amounts. For close matches - from parents and siblings on out through 2nd cousins - the quantity of shared DNA is high enough that there is no doubt about a family relation.

Smaller match amounts are more complicated because they can arise from a wealth of statistical possibilities - anything from a 3rd cousin once removed to an 8th cousin or something in between. Anything over a range of several generations is possible, as well as half relations from sharing only a single common ancestor - for example, the great-great grandmother who was widowed, remarried and had more children.

The goal for someone building a family tree is to trace a shared match with high confidence to a Most Recent Common Ancestor (MRCA), usually a set of ancestral grandparents of some degree of greatness. And ultimately, we're trying to identify previously unknown ancestors in order to build our tree farther back. A single DNA tester who matches us can only indicate that a relationship exists, and based on the match quantity we can make some predictions about the closeness of that relationship. In order to identify genealogical connections, we need to know more about the people

YOUR RELATIONSHIPS
How You Connect to Various Family

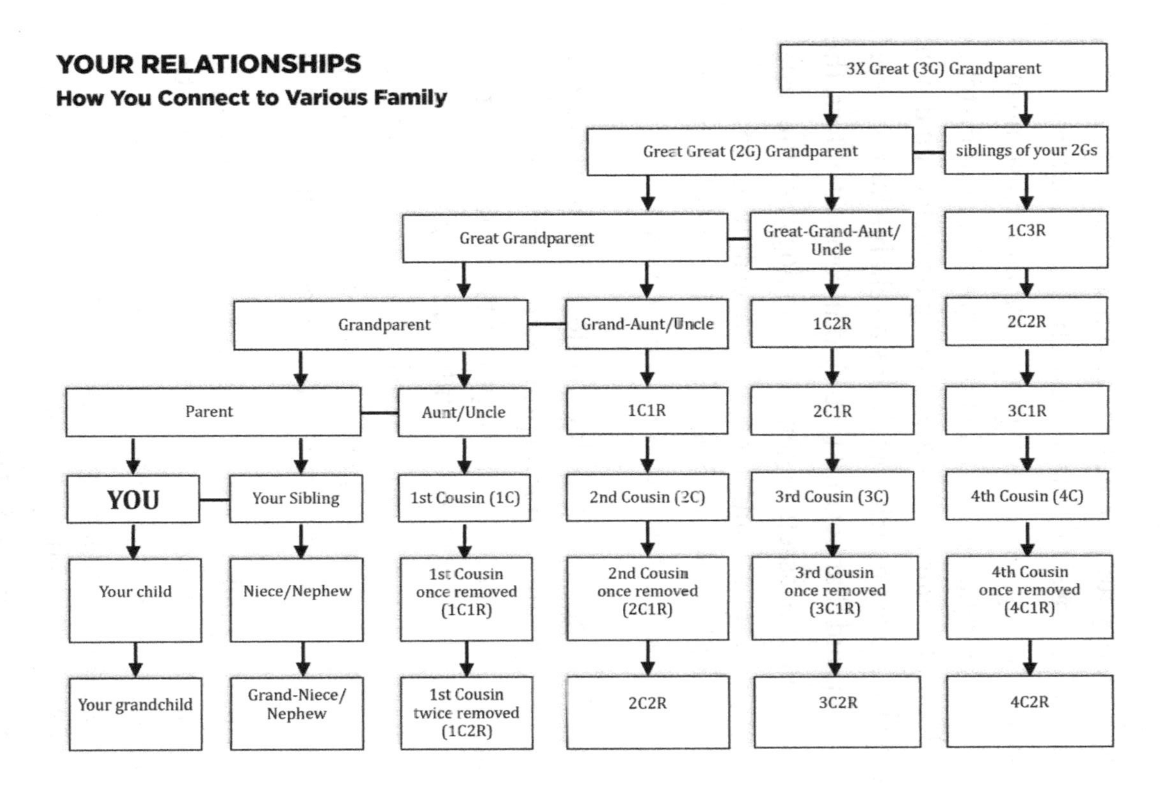

who share matches with us and that tester.

For each of your matches at the big testing companies, you will find a set of shared matches with that person. Distant "unknown" relatives may also match some of your close relatives, and this triangle of shared matches may point toward a common family branch even if none of your shared matches includes a reliable family tree. It is often said that beyond testing our revered elders, the most valuable assets in genetic genealogy are the tests of known 2nd cousins, who in theory allow us to determine how an unknown match is related to one of four sets of great-grandparents in some way.

It is often said that DNA doesn't lie, but that trivializes the work required to interpret the evidence. Close relations like parents, siblings, aunts and uncles, first cousins, etc. are essentially indisputable given the predicted amount of shared DNA. In the case of an unexpected parentage, those close relations may only match half as much as predicted or not at all.

Due to the randomness of recombination, matching segments of autosomal DNA are reliable to the genetic genealogist typically over five to seven generations, corresponding to a range of perhaps 125 to 250 years into the past. Testing elderly relatives allows us to reach farther back into the past.

Inheritance is more random and unequal from more distant ancestors, as some of those segments disappear in just a few generations. Others may persist for far longer. We don't inherit the same amounts of autosomal DNA from each of our 16 great-great grandparents (2Gs). As we go farther back through successive generations, we find that we haven't inherited any identifiable segments from many different lines in our family, while segments from a handful of other family branches may still show up 10 to 12 generations distant.

Your 3rd cousin (3C) is five generations removed from your shared set of 2G grandparents. One in every ten of your 3Cs likely won't share ANY common segments from those 2Gs with you. Because siblings don't get the same 50% of their DNA from each parent, it is quite common to match some siblings in a family of 3Cs, but not others. This is why testing multiple siblings and close cousins is invaluable in researching your family's genetic tree.

But the money-maker element of these tests that gets more of the attention and more of the advertising budget is the ethnicity estimates. Apparently, the marketing folks for the big testing companies have discovered the general public is more interested in knowing our ethnic composition than our actual ancestors.

Of course these estimates can be a bit controversial. Beyond the 5 to 7 generation range we simply don't have any DNA from some of our ancestors, so by default they can't be included in our deeper ancestry. Nonetheless, the major testing companies continue developing proprietary algorithms and methods to offer some hypotheses about where our genetic ancestors were farther back in time, perhaps 300 to 1,000 years ago. As the testing database gets larger with more well-verified family trees, there are certain genetic signatures that appear common to various historic populations around the world.

These estimates of ancestral ethnicity are probably most accurate for ancestors who came from known insular or isolated populations. Over the past few years my mother's estimates at Ancestry.com have been essentially two-thirds Irish and a third English. As more people with known connections to places in County Cork have added to their DNA testing database, the algorithms recognize that Mom shares genetic signatures with people who have known ancestry in both Southwest Cork and East Cork. Naturally, the database of these genetic signatures is not expanding at the same rate for all populations around the world.

And of course, just as with our genetic ancestors, we often find some surprises in our ethnicity estimates that don't jive with what we've been told. For instance, Mom's estimate from MyHeritage is 70% English, 17% Irish and 8% Scandanavian, while at FamilyTreeDNA they predict her makeup is 57% from the British Isles,

31% from western Europe, and 11% from Iberia (Spain and Portugal). We have no real way to know which company is correct, even for the ancestral DNA she did inherit.

For an American, who embraced their Irish heritage as family tradition all their lives, to find the bulk of their ethnicity appears to come from Scandinavia can be a bit of a shock. At the very least, it could suggest a shift in holiday meal planning from corned beef and cabbage to lutefisk.

DNA matching provides us evidence, not answers. A well-documented and validated trail of records and research must be used together with DNA evidence. The farther into the past we go, the less certainty we can have about our findings. As an added bonus, anyone with even a modest amount of experience in sifting through their DNA matches will tell you that our ancestors often kept secrets. Sometimes it turns out that a family legend was simply a great story. Other times it turns out there might have been some truth to it. And once in a while, we find that Grandpa was never actually our biological Grandpa at all.

Just another day exploring Pandora's Box, right?

YOUR TREASURE: You may be reluctant to do a DNA test for a variety of reasons, and many of those are legitimate concerns. And while I can't promise you'll learn a great deal more, you are likely to learn a lot in a short period of time. If you test at Ancestry or 23andMe, you can upload your DNA at other testing company sites and cast a wider "cousin net" of matches in their databases as well. And you get the added amusement of comparing your ethnicity estimates between them.

CHAPTER 7
Two Tubes of Spit

My parents' DNA tests led to a whole new world and an unimaginably epic adventure.

While I had several pieces of paper and a lot of family stories, along with a few odd mementos and some pictures, this journey really careened into the wonderland world of the "you can't make this stuff up" with two simple tubes of spit.

I've been lucky. A lot of my ancestors, great-great and 3G grandparents in particular, were pretty easy to trace. Thanks to my grandmother's self-assumed job of keeping her family's history, there was a fair amount written down already. I started probably knowing half of my 32 3G grandparents. And at this writing, I pretty solidly know all 16 on my dad's side and at least 11 on my mom's. They came predominantly from English-speaking countries, and before 1900. Most of them wound up coming to Connecticut, a state that keeps records pretty well, given its long colonial history.

While Gram had given me a huge head start with her quarter of my ancestry, I did have the basics of three or four generations in most of the branches of my family. My research toggled between being overwhelmed by her roots, because seemingly millions of people also had my grandmother's colonial heritage and traced it back to England in the 1400s and beyond, and stymied by the brick walls for my other three grandparents in much more recent Canada, Ireland and Scotland.

I learned about DNA tests for family ancestry sometime in the summer of 2015. I don't even remember exactly how it came up. My friend Lea was excited about tracking his Coryell family ancestry through a specific test of the male gender chromosome, better known as the paternal chromosome or y-DNA. y-DNA tests allow us to reach farther back into the past than the common autosomal DNA tests. But it is only for a single line of male descendancy - the patrilineal or father’s father's line from which our surname would come, in theory.

I guess that my dad had some curiosity in learning more about the McKnights. Aside from the intrigue of his father's family trying to get as far away from each other as they could, there was that bit that we were of Clan MacNaughton, and how at some point we'd been exiled from our family lands in Scotland for sheep-rustling or some other such legend. So Dad took the test, and joined a y-DNA project with over one hundred other McKnight men, some claiming a provable known lineage back to one of the earliest Clan MacNaughton chieftains.

And we matched....none of them. Zero. Zippo. Zilch. Among all these various McKnights of Scotland and Ireland, the Scots-Irish pioneers who fled west across America until their pursuers tired of chasing them, we matched nary a one. As of this writing over 150 McKnight men have joined the project, and our record remains a perfect goose egg.

What that implies is that we have no common ancestor with any of them in probably the last 500 years or so, and that raised all of our eyebrows. "We're not McKnights? Well, what are we then?"

My grandmother played another huge role in my life, too. Mom lost both of her parents by the time she came of age. So Gram became like a surrogate mother to her, her biggest champion and her role model as a mother-in-law and a grandmother. And of course, by the absence of Mom's parents, it meant that Dad's family history played an outsized role in my childhood. Now it would seem a big piece of that history just got murky.

By this point I learned about the autosomal DNA tests from Ancestry.com. Both parents could take them. Mom had a natural curiosity to know more of her family history, based on stories she'd heard from her parents and grandparents. And now Dad had incentive. We ordered the tests.

None of us were quite prepared for the sheer volume of people who matched one parent or the other. They each suddenly had literally thousands of "new" cousins, most with no apparent family tree connection. However, even the few dozen who did obviously connect to a shared ancestor opened up a Pandora's box, which continues to frequently sprout interesting and occasionally exciting new discoveries. The rest of the matches were like someone had just dumped a 30,000 piece, three-dimensional puzzle on my desk.

My parents taking the DNA test launched us a generation farther into the past, right out of the gate. Since they were born in the 1930s, DNA evidence in theory might extend our family research back in time as much as 250 years. In other words, we had a chance of finding connections to ancestors that had been born well before the Declaration of Independence was signed. Since Gram had all that colonial ancestry, it was likely that a) some of Dad's matches shared some AND b) we might figure out the common ancestor with some of them.

As for the other three grandparents' branches of the family, it was a daunting notion. Thousands of DNA matches, all likely having inherited their DNA from some common ancestor we might never identify - particularly those ancestors in Ireland.

Ah, Ireland. Mom's mysterious big family of cousins, all carefully typed out and tracked in my Great Aunt Muriel's sheets of family heritage, all starting from people with mysterious origins in Ireland.

Ireland would offer us many surprises, and wasted little time in doing so. On my father's father's father's grave in my Dad's hometown, the stone reads simply Andrew McKnight, "Here Lies a Proud Scot." Indeed, the records show he was born in Scotland. A little digging on Ancestry combined with DNA matching illuminated a new aspect to his story - he was born mere months after his family emigrated across the Irish Channel from County Down.

And now my dad, who has spent most of his life avoiding his mother's careful collection, curation and dissemination of family history, is suddenly interested in his heritage. For he has learned that we may not be McKnights, and in fact have roots in Ireland rather than Scotland, and the bedrock of his father's heritage has now morphed into rather loose gravel. Before long, he is charging his two adult children with planning "The Last Great Family Adventure." Just the four of us, with no kids, spouses or pets, visiting the lands of our ancestors - in a tiny minivan, driving on the left side of the road. What could possibly go wrong?

YOUR TREASURE: Remember that family tree I suggested that you start back at the end of Chapter 2? If you take a DNA test, here lies the real power of the bigger testing companies. Fill in your tree and connect it to your DNA test results, and you will soon get suggestions about how you are related to your matches who also have their own family trees. While some of these hints will be plainly incorrect, it is also likely that some of your DNA matches who have carefully researched their trees might have a new answer or two for yours. Take the time to carefully vet your possible connection to each new ancestor suggestion.

CHAPTER 8
Legends and Myths

There is often some truth in family stories, but they can be hard or impossible to verify. Those fantastic stories do give some incentive to further explore our family history, and sometimes what we learn is even more amazing then the legend.

Most people have some family story that has been handed down, either accepted as gospel truth or told with a chuckle and a wink implying some skepticism. Those tales that Gram told us about being descendants of Robin Hood stick out for me. Most every family has its narratives, and many of them include myths and legends that are nearly impossible to verify. And as with most myths and legends, there is probably some kernel of truth somewhere within. But what is that truth and how much has been obscured and reshaped in the countless tellings and retellings?

Those tales might be best described as identity totems. They are part of what we hold in the fabric of our family memory and story. They help us in how we describe ourselves. And as we dive into the deeper research of genealogy, we run the risk of unraveling that tapestry, even as we weave in new colors and textures. It is important to realize that we are still who we are, regardless of whether those stories are 100% accurate. Learning that an ancestor was not all they were cracked up to be, didn't do the things they said they did, or in fact wasn't an ancestor at all can be a bit traumatic. But it does not change who we are, nor does it untwist any of the DNA with which we came into this world. And once in a great while, we find that we might actually be descended from Robin Hood.

The keepers of those family stories are usually the elders. So in 2004, I got the idea to film two of my great aunts talking about family history, and I was lucky to have one on each side of Mom's family who were willing. My great Aunt Muriel (97 and going strong as of this writing!) dutifully kept the family history that her mother handed down to her of Mom's Murray and Ryan maternal side. And Mom's father's youngest sister, my great Aunt Phyllis, had similarly been the keeper of the Bagley-Baker family stories, about which I'd known even less.

It was a fascinating and delightful afternoon, full of Aunt Muriel's endless supply of anecdotes that I have still not exhausted. And while Aunt Phyllis had been the youngest of Mom's father's siblings, she had a few fascinating tales to tell too, including how the family fortune went down with the Titanic!

Humans have a fascination with disaster stories, and discovering a family connection to one feels like getting a small merit badge in genealogical notoriety. Few things get attention in casual family history conversations quite like "My ancestor went down on the Titanic." It is one thing to have a family legend, and entirely another to support it with evidence. And surely any moderately interested party is going to try to verify a Titanic story.

Mom's 2X great grandmother Leavinia Kirkland and her brother Charles were born in the 1840s in maritime Canada. The family origins remain murky. Supposedly their parents emigrated from Scotland in the 1820s and established themselves as silk traders in northeastern New Brunswick, possibly in connection with family back in Scotland as suppliers. Charles Kirkland turned away from the family business, caught up instead in a wave of Free Will Baptist fervor sweeping through the region in the mid-1800s. He had become a preacher, serving whatever communities in Maine and New Brunswick would pay him. By his sixties, he seems to have largely been a traveling preacher, apparently blessed with some skill at switching denominational style and delivery, as many as three times in a weekend.

In the autumn of 1911, he somehow got word that his two heirless uncles in Scotland had passed away and that there was an estate needing legal settlement. As his parents' eldest surviving son, he made the trip across the sea to handle the family business. After several weeks, he wrote a letter home to his daughter Maude expressing his frustration at not locating anyone or anything related to the estate and despaired of being stuck in gloomy Scottish weather until April. He mentioned having spent $500 already on the excursion, and that he had to be careful to have enough to get home in the likely event he'd been chasing wild geese.

But apparently he DID find something by April of 1912. It is hard to imagine that an itinerant preacher's remaining pocket change would be sufficient to buy a second-class ticket on the gleaming new first in its class trans-Atlantic steamship. But records show that's exactly what he did. It cost him his life, and perhaps his heirs a considerable bit as well. His body was never found, and whatever papers and secrets might have been traveling with him went down with the good ship RMS Titanic.

The pieces of this long-held family story may be partially or completely true. Many of the details are difficult or impossible to validate, though a great many people have tried. What we do know is that we have plenty of DNA connections that share his and Leavinia's parents as common ancestors. And people remain fascinated by any Titanic story to this day.

I wear my grandfather's name, as he wore his father's, and his father wore his father's. It was bound to happen, that someday I'd figure out I'm the fourth Andrew McKnight in five generations. To say the Scots aren't particularly imaginative with naming their children might be giving them too much credit. Downright boring and repetitive, as well as confounding to genealogists, the Scots have seemed to recycle the same dozen or so names endlessly through the centuries. The fruits and roots alike in my grandfather's family tree are repeatedly adorned with Margarets, Sarahs, Jameses, and hell yes, Andrews. The constant repetition forced me to identify my Scots relatives by a de facto title – my great-grandfather Andrew the Fiddler, his daughter Margaret the Opera Singer, and so on...

Someone someplace, likely my great-grandfather, hammered Scots-ness into my grandfather's family. It's sort of funny, given that his father brought the family from Northern Ireland to work in the booming mines of southwest Scotland around 1863. And so even though Andrew the Fiddler was born in Scotland, it is likely the family and their fellow Ulster Scots returning to their ancestral homeland in the Lowlands were instead regarded as lowly Irish. He may have clung to his Scots identity in self-defense, or perhaps it was an affirmation of ancient family origins.

Whatever it was, he certainly carried it to America with him in 1888. When he met and married Maggie Jane Robinson (born in Belfast, raised in Glasgow) in Massachusetts, they left no doubt in any of their American kin that they were Scots, and damn proud of it. They clearly raised their eight children born in America resolute in their conviction that they were Scots too.

Those children and their children all heard that we McKnights were of Clan MacNaughton. As a child I was given the book *Clans and Tartans* by Gram's friend, Colonel Frederick Manross, which he kindly inscribed from "one Scot to another." I still have that book, and in it recently I found two index cards, written to me by my grandfather - detailing that we were of Clan MacNaughton and Clan Robertson (from Maggie Jane's family).

But my older cousins had also heard family stories connecting us to the "Clan of the Black Douglas." I had no idea what the hell that term meant, how it fit, or if it implied that we descended from dangerous outcasts of some sort. Then there was the long-told family legend that our ancestors had been run out of Scotland for sheep rustling (how novel!) Presumably this was how we wound up in Northern Ireland, before "triumphantly returning" to the homeland in the 1860s and 70s.

My great grandfather himself is largely responsible for this confusion. While they enthusiastically handed down the Clan MacNaughton legend, there are plenty of other confusing bits of evidence. Particularly puzzling is the letter my grandfather received after his father Andrew the Fiddler's death, from a Boston-area chapter of Clan Wallace, as in the Scottish legend William Wallace - a.k.a. Braveheart, expressing condolences and mourning "the loss of a Senior Clansman and Senior Past Chief."

There's also that small matter of Dad's paternal DNA test, his y-DNA which traces only the line of father's father's fathers. While our autosomal DNA connects us neatly to all of the other McKnights in our line going backward to our early 1800s genealogical brick wall in County Down, that ancestral patrilineal lineage is another matter entirely. Since we do not match a single other McKnight, perhaps our ancestors stole more than sheep - maybe we swiped the name on the way into Ireland as well.

Four different Clan origin possibilities, ancestral exile and sheep rustling. Or, just another day in the life of dusting off family legends for a closer examination.

> ***YOUR TREASURE: Does your family keep any legends or fascinating ancestories? It is vital and precious to film your elderly relatives while they are still here to tell the stories they heard as children. Sometimes the tale is enough to spur us into doing more research, but what we find might***

turn out to be even more interesting. And an old adage is always worth keeping close whenever a new record or story falls into your lap - "Trust, but verify."

CHAPTER 9
Meeting Aretas Culver

How my previously unknown 3X great grandfather's story has intertwined with mine in these years turns out to be but another chapter in the long ties between his commanding officer's family and ours.

As often happens these days, I met my 3G grandfather Aretas Culver online.

My grandmother has this big beautiful bush of a family tree, tracing many of her myriad branches back to 1500s England. Yet in the middle of all her rich heritage, three stumps simply vanish in the midst of rapidly growing Connecticut in the early 1800s.

As one might guess, I have spent countless hours chipping away at these stumps. I had clues enough, but they led nowhere. In the case of her great-grandfather

Culver's line, I suspected the handwriting of a nine-year-old on a DAR lineage sheet may have been the problem. There was a Culver in one of Gram's Revolutionary ancestor's lines, but his first name made no sense.

It was Christmas 2014, and as always, we returned to my childhood home in eastern Connecticut. Much like some species of fish return from the ocean to the upstream pools of their birth, my sister and I from opposite directions load up our families and make the journey of many hours to spend each Christmas with our parents in their tiny house. We've grown to cherish that at our ages we still can do this, even though the traveling is hard.

Mom and Dad have TVs in every room, and they don't suffer from a lack of use. The users suffer a bit from lack of hearing though, so there are usually at least a couple TVs competing for one's attention. If they are not set on different channels, they are slightly enough out of sync to drive a normal hearing person batty.

So you pick a room and plant yourself in it to keep from losing your marbles. Often I'll pick the living room because the couch is comfortable, and amuse myself on the laptop if I'm not all that interested in what they are watching.

For some reason on this particular evening, I was noodling about looking online for cracks in our Culver brick wall. I admit I was a little cranky about not learning anything after several hours of work spread over a few weeks. I decided to revisit spellings of his strange-looking first name. I tried a few things and finally tried A-r-e-t-u-s, and B-I-N-G-O!

There he was, staring back at me from my laptop, my great-great-great-grandfather Aretas Culver. (It is quite common to find your ancestor's name spelled more than one way, and to never be able to confirm the spelling intended by those who named them.) It was astonishing enough to actually meet him "face to face." But then I realized his picture was coming to me through the Connecticut Public Radio website. It was the lead image to a story they had done six months earlier about an exhibition at the Connecticut State Library, featuring photographs of the state's Civil War veterans before they had been imprisoned at Andersonville, the notorious Confederate prisoner of war camp.

I shuddered for a moment. Simply mention Andersonville to anyone who knows anything about the Civil War, and it evokes emaciated, diseased and dying men, with their jailers faring not much better. And now an arc of my story went directly through that hell:

> "Camp Sumter was only in operation for fourteen months, however, during that time 45,000 Union soldiers were imprisoned there, and nearly 13,000 died from disease, poor sanitation, malnutrition, overcrowding, or exposure.

The first prisoners arrived (in Andersonville) in late February 1864. Over the course of the next few months approximately 400 prisoners arrived daily. By June 1864 over 26,000 prisoners were confined in a stockade designed to house 10,000. The largest number of prisoners held at one time was 33,000 in August, 1864.

The Confederate government was unable to provide the prisoners with adequate housing, food, clothing, and medical care. Due to the terrible conditions, prisoners suffered greatly and a high mortality rate ensued.

Approximately 19 feet inside of the stockade wall was the 'deadline,' which the prisoners were not allowed to cross. If a prisoner stepped over the deadline, the guards in the 'pigeon roosts,' which were roughly thirty yards apart were allowed to shoot them."*

**Courtesy: National Park Service*

It turns out that the picture and the public radio story were only the top return on my search. Next up was even more interesting - an excerpt from a personal letter Aretas had written that was published in the *Hartford Evening Press* on Sept. 22nd, 1862. In it he described the experiences of his unit, the 16th Connecticut Infantry, during their final major action at the Battle of Antietam:

> "The 16th were posted on the extreme left of Burnside's column...The enemy commenced a flanking movement, which they were nearly successful in accomplishing, having treacherously displayed the U.S. colors until ready to grapple with the foe...At that point of attack, they lowered their standards and hoisted the rebel ensign in its place; then it was the Acting Brig. Gen. Harland ordered Col. Beach to change position to escape their deadly cross-fire. Col. B attempted it, but in vain, and he was obliged to report to his superior that his regiment had never had a battalion drill, but one dress parade and scarcely knew how to form in a line of battle."

His letter was included in the book *A Broken Regiment*, written by Dr. Lesley Gordon and published five weeks prior to me arriving at my parents' house for Christmas. Within a few minutes search, I had stared my ancestor in the face. I learned that he had been in the battle at

Antietam with the greenest of new recruits, who failed in the face of the furious charge, which probably extended the Civil War for three more years, and then wound up imprisoned under the most unimaginable horrors by war's end.

Such a detailed narrative of a regiment, whose history was bookmarked by colossal failings, is rare among Civil War books. The inclusion of Aretas Culver's own words, coupled with his easily found photo, instantly seared him into my consciousness. I had solved the mystery of my 3G grandfather alright. But he had a story. And it had just become ours.

It was not difficult to trace the whereabouts of his regiment online. I learned that he signed up in the summer of 1862, answering Abe Lincoln's call for volunteers and claiming the bounty payment offered. He and his townsmen formed Company K of the 16th Connecticut, were shipped off to Ft. Ward near Washington, DC, and issued old Belgian rifles. They had one drill with those relic weapons before they were rushed aboard a train to join General George McClellan's Union Army, which was marching to intercept Lee's Army of the Potomac and their intended invasion of the North.

Of course, green troops like this weren't supposed to see action at Antietam. One could offer several plausible reasons why the Union Army moved so slowly at critical times, or criticism of strategy and performance in the field. The fact is that once Union General Ambrose Burnside's army finally took the bridge that now bears his name, and began marching on what remained of Robert E. Lee's battered and exhausted Confederate Army, they were several hours behind schedule. Emergency reinforcements were the only ones available. And thus these raw recruits wound up guarding the left flank as Gen. Burnside's army marched through the hilly cornfields, pushing Lee's weary and depleted army back toward the Potomac River.

The 16th Connecticut will forever be a footnote to what military historians generally agree is one of the "greatest forced marches in military history." Confederate General A.P. Hill's battle-hardened shock troops had captured the Federal Armory at Harpers Ferry two days earlier. When Lee's urgent plea for reinforcements arrived at 6 a.m., they responded by covering the 17 miles in just eight hours and arriving on the battlefield in unbroken stride. Their lead units flew the Stars and Stripes and wore blue Union kepis, further confusing their identity and intent to the unsuspecting Union soldiers. Their arrival saved the day for the Confederacy, resulting in a stalemate on the field and perhaps allowing the bloody war to continue far longer.

The 16th understandably had cracked and fled in the face of such a determined assault. They were blamed for failure at a key moment. They were accused of cowardice. After digging graves at Antietam, they were marched under the guard of their own army, for concern that they would "skedaddle." And much to their humiliation, ill rumors of their conduct under battle made their way back home.

They would see grave detail again at the bloody Battle of Fredericksburg in December and then spent a year of relatively comfortable and uneventful garrison duty in Tidewater Virginia, far from the horrors and relentless misery of war. Then in early 1864, their routine barracks life ended abruptly when they were shipped off to North Carolina to resume active duty. They were assigned to Plymouth, near the mouth of the Roanoke River on Albemarle Sound.

The town and its garrison were surrounded by the Confederate army on April 17th. After enduring four days of shelling and worsening conditions, they were forced to surrender on April 20th. The captured men were shipped by train to Andersonville. In late November, Aretas was paroled (released in a prisoner exchange) and sent to Savannah after over five months as a POW. Suffering from scurvy, he returned home to Bristol on a 30-day furlough on the 4th of December, but never recovered from the effects of his captivity. Aretas Culver died on Feb. 9, 1865, leaving four children from his deceased first wife, and his second wife and their young daughter.

I spent a lot of time piecing together the family over the next few months. I pored over my copy of *A Broken*

Regiment, mesmerized by the stories of these young men who signed up to keep the Union together. They were failed by their commanders, and in turn, failed tragically themselves.

I later learned that en-route from Antietam to Fredericksburg in late October, Aretas had most likely spent a night not two full miles from our village of Lincoln, Virginia. He and about 30,000 other Union troops camped in the nearby town of Purcellville, in the woods around what is now the Babe Ruth League baseball field.

My dad knew nothing of this story. Aretas joined the army at 42 years of age, already a remarried widower raising four children from his first marriage and an infant daughter at home. By then the failures of the Union army on Virginia battlefields from Ball's Bluff to Bull Run had become a national embarrassment. It would be logical to think he would be clearer-eyed about his commitment and the potential consequences than the typical young enlistee seeking adventure, notoriety or simply escape from daily drudgery. While from the comfort of the 21st-century I would love to ascribe his motivation to something noble like ending slavery or preserving the union, he may well just have needed the money. His motivation for going to war likely will forever remain a mystery,

His only son, Wilson, was Gram's grandfather. He was 16 at the time of his father's death. He was married and divorced three times on account of being "habitually intemperate." So I assumed that Gram knew little of Aretas because Wilson had long been out of her mother's life. It was only many years later I found a receipt for the burial of Wilson Culver in her family history album, paid for by Gram's family, alongside Aretas' official notice of promotion to Sergeant in 1863.

Gradually though, a longer and deeper thread emerged. The 16th was led by Captain Newton Manross, who was killed at Antietam. Aretas Culver helped escort his body home to Bristol. He likely wrote his letter on the train ride back to Connecticut, while his comrades dug graves under the careful and perhaps contemptuous watch of their fellow Union soldiers.

My grandmother Madeleine was close friends with Betty and Fred Manross, the grandson of Newton. Their family was wealthy enough to do considerable philanthropic work in their hometown, including building a new library across the street from my grandparents' house.

A couple of years after discovering Aretas and his story, my cousin Marjorie gave me a different compendium about Bristol history, *In the Olden Time New Cambridge*, which fascinated me for weeks. At the time of its writing in 1907, only four families remained in Bristol that had been there during its founding years, 1721-42. The

Manross family was one. And the Jerome family, ancestors of Aretas Culver's mother, were another. I realized then that our families have shared over 200 years of history together. My cousin's family still lives in Bristol, which means that our family and the Manrosses have been there since the town was founded. My grandmother and her friends Fred and Betty were simply carrying on a tradition of sorts between their families, whether they knew it or not.

And those long woven threads of kinship were one of the many strange and amazing things one may find in the treasure chest.

> ***YOUR TREASURE: There are often additional resources to learn about ancestors who served in wars. The US National Archives (NARA) in Washington, DC, is the main repository for American records, and these collections are available online (usually for a fee through one of the various family research portals like Ancestry or Fold3). While there is some paperwork to be filed in advance, you can actually schedule a visit to the Archives to personally examine the physical records you've located online. There is no cost to visit in person, but the experience of holding your ancestor's records in your hands is priceless. Avoid photocopying fees by taking an iPad/tablet for imaging records, and before leaving home be sure to clean the lens well.***

CHAPTER 10
By The Time She Got to Woodstock

Finishing a childhood journey that my mother's family started.

I never knew my mother's parents. Her dad died when she was 13, and her mom was already dying from cancer at my parents' wedding. Before the year was over, Grandma Hazel was gone. Before she turned 21, Mom stepped into a new world of marriage and finishing college, while moving to eastern Connecticut's rural hills - a full hour away from her parents' families.

Thus, my dad's parents played an outsized role in all our lives. They frequently liked to get out of the city and drive the two-lane roads through the rural hills and valleys of Eastern Connecticut to visit us. I was close to my grandparents. And my grandmother, in particular, adopted Mom as her own. Gram was a steadfast champion who valued her daughter-in-law's worth and work. In her way, she did whatever she could to make Mom's life a little easier, all the while with cheerful and witty energy, and her before-dinner highball.

Outside of Mom's sister's family, her relatives didn't visit much, and we didn't go there often either. They all lived fairly close to each other south of Hartford. They lived in the Connecticut River Valley towns where her mother's Irish kin had come from New York City in the 1850s and 60s, and where her father's family arrived from western New Brunswick during the Roaring Twenties.

Both families remained a confusing mystery to me well into my adulthood. I only had five cousins on my mom's side, and two of them I'd only seen a handful of times since my uncle was in the Air Force and constantly being relocated. The idea that Mom had over 20 cousins on her mom's side alone was too much to process. And without having much prolonged contact, there wasn't a lot of impetus to figure it out. I'd get curious and ask now and again, and add some bit of new understanding that might stick with me for a while....or not.

Mom retired from teaching in 1998, and my dad looked forward to having her all to himself. Much to his chagrin, she quickly bloomed into a "new" life that included reconnecting with a lot of her family. Free to travel on any given weekday and to make plans with or without him, she did and continues to do so. And as one-by-one her aunts and uncles passed away, I got caught up in wanting to know more of this tangle of my family roots.

Over the course of time, I pieced together a lot more details of how Mom's paternal ancestors had likely been Loyalists fleeing after the Revolution to British Canada to start over, and Irish immigrants arriving in the 1820s and 30s looking for a new life. They were among the settlers who helped found Woodstock, western New Brunswick's first incorporated town in 1856. They had big families both in town, and on farms and small rural settlements in the surrounding hills. My grandfather Perley was born there in Woodstock in 1907. In a wave of people from along the Maine-New Brunswick borderlands migrating to southern New England to fill an abundance of jobs, his parents and their family moved to the Hartford area around 1920.

When Mom was 11, the family packed into their Graham sedan to go to Canada to visit her dad's family for the first time. But Perley was already feeling the effects of the illness that would claim his life a few months later. A couple of hours into the trip, he felt too sick to continue, and so they returned home. After he passed, the opportunity never came up again. Mom grew up, settled down, and her only contact with her dad's family was rare and short visits with his youngest sister Phyllis.

As I learned more about the family's roots and branches, I got the itch to see Woodstock - a journey of about 850 miles from my Virginia home. But while the typical traveling troubadour lifestyle (including income!) doesn't allow for much recreational travel, it does facilitate staying in touch with family and friends offering safe harbor in myriad corners of the continent. My sister's family lives near Bangor, Maine, just a couple hours south of Woodstock. And each August for several years I served as Musician of the Week for a retreat center on the southern Maine coast. True to form, work made it possible and safe houses on the way made it feasible. Most importantly, late in the spring of 2015, I talked Mom into planning the pilgrimage.

Woodstock sits just 10 miles into Canada at the true end of I-95, the highway that runs from south Florida to the eastern border with Canada, and then was extended by the Canadians to its terminus at Woodstock. We figured with three of us to share the driving, and the long days of summer, we'd be able to spend a full day exploring and still make it back to my sister's house that evening.

While we knew well in advance that passports would be required and had taken the requisite steps to get ours updated, mine had not arrived by the time we left Virginia. And so we rendezvoused the day before our trip with the sobering realization that our plan had a very real Achilles heel. We figured - and hoped! - we could talk our way across the border, as I couldn't imagine anyone better able to officially vouch for me than my mother and my wife!

The workday bustle of Bangor fell quickly behind, giving way to the north woods and occasional paper mill or lumber town. Mount Katahdin, Maine's most famous natural landmark and the terminus of the Appalachian Trail, loomed off to the northwest. Despite being at the northern end of the busiest interstate in America, the Border and Customs station was not overly busy. My missing passport was greeted with only modest suspicion and they waved us through quickly, but not before assuring us that our country had to take back its own citizens and that we would be able to get home.

We passed those last few miles into Woodstock a bit like kids in a candy store - a new country, a new experience. We snapped a few pictures of the rolling forests and farmland, and passed signs for Richmond and the Campbell Settlement, two names that had appeared several times in Mom's family history. Anyone who has traveled along the 1,900 mile length of I-95, or experienced the oddity of seeing the sign at its southern terminus in Miami, might well know the weird sensation when we crested the hill and passed "Highway 95 Ends In 2 Kilometers."

Woodstock is a handsome town, surrounded by the hills around where the Meduxnekeag River dumps into the St. John's. The town center is near the confluence, where the brick facade of Main Street shops, the iconic

Canadian Tim Horton's eatery, and the town library formed the bulk of the business district.

While our trip had many purposes, it was heavily weighted with emotional significance despite our complete unfamiliarity. We had no living relatives to meet, no family homesites to visit, but it was still my first trip to a place where my own roots touched down. Everything about it felt strange and exotic, enhanced by remarkable summer days of ever-shifting cloud and sun interspersed, giving a dramatic and constantly changing lens to our experience. And after a valuable orientation session to collect some family history and local guidance from the historian at the Woodstock Library, we were ready to chase some ghosts.

Nortondale had been a small farming and timber community some 20 miles east of Woodstock that gradually came to life in the 1880s. Mom's great-great-grandfather owned land and raised a family. there. Our historian had given us some records from the primary school that Mom's grandmother and her siblings had attended before the family relocated to Woodstock.

Our route crossed highland fields with sweeping vistas of western New Brunswick, with but a few small clusters of houses and the occasional rural commerce to serve up machinery repairs to remind us the land was still inhabited. The sky kept up a constant aura of surrealism, with a handful of sporadic raindrops dancing between brilliant columns of sunburst and shadow. We passed an old hilltop cemetery as we approached Nortondale on our iPhone map.

But when our blue dot arrived at our destination we found essentially....nothing. A farmhouse with fields around to one side, the T-intersection as it appeared on the map, woods here and there, and a junkyard full of old cars just in view down the road. There was no trace of the settlement or the people who had lived there, save for the old cemetery. In contrast to the romanticism associated with the relics of "ghost towns" in the American West, whatever once was here seemingly vanished into thin air and thick forest.

Four miles down the road, we were glad to find the small community of Temperance Vale still alive in the hills and fields around the crossroad, including three churches and their cemeteries. Our same family roots stretched here too, so we spent some time walking about the small graveyards fruitlessly searching for family names.

Nonetheless, we had indeed "touched ground" in places walked by our ancestral family. A whole set of chapters of my mother's story were written and turned in these hills, even if we couldn't directly connect to any visceral part of it. I was deeply moved to be where they had lived their lives, gone through the moments of humor and sadness in their daily routines, and struggled with the hardships of the times and the long Canadian winters. Our ride back along the bank of the St. John's was largely animated with our musings and conversations about what their lives might have been like. And as we passed houses, businesses and communities of current residents, we wondered how different Mom's life might have turned out had her father's family stayed in New Brunswick. Their emigration was essential to my existence.

When we returned to Woodstock, we went up to the hilltop cemetery where we'd been told three of mom's great-grandparents were buried. Within about a half-hour and a bit more walking, we found all three stones and lingered briefly at each. It was almost as if simply by staring really hard at the etchings on a granite slab, we could somehow bring home some tangible connection to these people who lived and died long before any of us were born. It wasn't hard to imagine that we were the first of the family to return to these graves in as much as a century. The late day light added to a deeply visceral yet mysterious moment for us to contemplate during our long drive back.

Our final view of our ancestors' homeland came through restaurant glass during dinner in town. The watercolor day played out with the glassy smooth river awash in cloud reflections, the palate rich with a full spectrum of sunset hues.

Just as the Canadian border guards had promised, we were indeed ushered through the Customs gate and back into our own country, albeit with a few Canadian mosquitoes as contraband. The car fell quiet in the black of night as the weariness overtook my fellow travelers. We had a lot more to digest than a meal after such a day. In a strange way I felt as though I had somehow fulfilled a promise made to an 11-year old girl a long time ago.

Somewhere in the deserted southbound darkness of rural I-95, I told my mother "You made it." She nodded her head, and said softly "We sure did."

> ***YOUR TREASURE: What experiences might await you in an "ancestral homeland?" Can you visit a place where your ancestors lived in person or virtually? It is fascinating to look at a town and the topography on Google Earth or YouTube videos, but consider joining an online community dedicated to that place's history. Even if you can't travel to the places where your ancestors lived, collaborating with people who live in those areas can be immensely valuable.***

CHAPTER 11
The Lost Tribe: Margaret's Family

My mysterious great-aunt, the actress and opera singer, and black sheep of the family. She was key to understanding more of my family's musical inheritance.

I've developed many impressions of my ancestors through an incomplete patchwork quilt of clues. While it is all we have, "forensic personality development" is a poor substitute for actually having known people. It is a natural inclination to seek or interpret information in a way that supports one's existing beliefs or theories, an inherent threat to objective research in any field. I can personally attest to how our preconceived notions and confirmation bias can lead family research astray by overlooking or misinterpreting evidence that doesn't neatly fit our theories about an ancestor.

Frankly, keeping an open mind about some of these people can be pretty hard, particularly when they've been linked to egregious behaviors. I keep a note on a card next to my desk with a simple, necessary and frequent reminder, "It is impossible to properly study people's lives without the context of the times in which they lived."

My great-grandparents Maggie Jane Robinson and Andrew McKnight emigrated separately to Massachusetts, and likely met here through some connections in the insular community of Scots emigrants in the port city of New Bedford. They moved around the Boston area after they got married in 1895. My grandfather was the third of their children, and as the first son he was creatively given his father's-father's name - Andrew. It is a good thing that names don't wear out, because the Scots would be reduced to simply numbering their children eons ago. I imagine that they lived in a tiny house, and by 1911 there were eight children living in it, from ages 15 to newborn.

In 1917 they moved to a new house in Bridgeport, Connecticut. I've seen the house online, as it was on the market a few years ago. It sold for only $50,000, likely due to a combination of size, age and neighborhood. I'd imagine the oldest daughters, Sarah and Margaret, wanted nothing to do with moving into a small house as young adults. My then 17-year old grandfather moved with them, so there were at least six kids living there. I imagine them having to sleep standing up to all fit in that house!

The great-grands probably raised their children the way that they were raised. Maybe because that's all they ever knew, or maybe they even resolved to not repeat the mistakes their parents made. The evidence suggests that they did a lot of damage, and it really doesn't matter much now whether it was by direct actions or inadvertent consequences. The three boys grew into full-fledged alcoholic adults, and most of the girls grew into a lifetime of frequently erratic behavior, including difficulty in maintaining healthy relationships with their own children. Whatever their reasons, my grandfather and most of his siblings went their separate ways as soon as they could leave.

Gramps next oldest sister, Margaret, was a complete mystery and a fascination to me. She was a wild one by reputation. I had heard she was a lovely singer and Great-Grandpa had sent her to Italy to study voice with some legend of the opera. My grandmother warned my cousins to never ask Gramps about her, which was odd and only added to our fascination.

During one of my tours across the deep South, my cousin Lee and I arranged to meet in person for the first time. Her grandmother Isabel and my grandfather were siblings who lived in the same town but didn't care much for one another. Their relationship or lack thereof was the "normal" in the McKnight family, rather than the exception.

Since Lee and I didn't see any reason to continue this pattern, we gladly took on the collaborative detective work to learn anything more about our shared roots. We worked for weeks before and after our meeting, slowly peeling back the onion of mystery around our McKnight family heritage. We were both intensely curious about the family we didn't know; especially our Great Aunt Margaret the Opera Singer, the second of the eight siblings. We knew that she'd married more than once, had some children and gone to the West Coast. That was about it. She clearly shared the McKnight musical genes, and we were eager to learn more of her story.

Slowly we teased out more details of her trail - a marriage in Massachusetts and a child in 1918, a marriage in Ohio in 1921, and a marriage and two children in California in 1928 and 1929. Apparently, she divorced again, and after the 1940 Census, the trail went cold.

In late 2015 I stumbled across two online newspaper clippings. There was a broadside from a 1929 *Reno Evening-Gazette* about an opera performance featuring Margaret and two other stars, noting she had studied with the Italian singing legend Fernando Carpi. The second, for a 1931 minstrel show, listed her in the supporting cast. A friend of mine pointed out that there may have been a stark difference in the career of a professional singer in the Roaring Twenties versus the Great Depression. It's a long way down from headlining the opera house to a small role in the minstrel barn.

On May 25, 2016, Lee messaged me that she was pretty sure she had found one of Margaret's great-granddaughters working as a chef in San Diego. She picked up the phone and called the restaurant, and within a few minutes was indeed connecting voice-to-voice with Sarah. They talked for a bit and Sarah passed along her father's contact info, saying he would be eager to talk with us.

And so that night I found myself on the phone with Margaret's oldest grandson Carson, who lives in Seattle. We talked for a good long while, obviously with lots to share and both realizing the magnitude of the moment. Although Carson had spent time with his grandmother and knew a lot of her story, he didn't know much about us at all. What had been a mystery all of our lives was suddenly right here, immediate and real. My hand cramped a dozen times as I furiously scribbled notes on my trusty yellow pad.

As a teenager, Margaret lost a Broadway role to Lillian Gish, reportedly in part because of her near-sightedness. She couldn't read the cue cards. Since Gish was only cast in one Broadway show before 1930, it must have been the role of Morganie in "A Good Little Devil" in 1913. The trajectory of Gish's career would eventually earn her the moniker "First Lady of American Cinema," beginning with the seminal silent pic (and Southern revisionist history) "Birth of a Nation" in 1915. Aunt Margaret's career was headed in a different direction.

Carson's mother Martha Housen was the first of Margaret's two California daughters, and she too was an opera singer. Like her mother, Martha had several short-lived marriages and a child from each of four of them. Her younger sister Jane had been a modestly successful west coast-style jazz singer. She too had a few children and led a bit of the gypsy life, as was common in southern California in the late 1960s. Despite their unsettled upbringing, music and theater were huge threads in the lives of Margaret's family too. Carson was close to his half siblings and some of his cousins too.

Later that night Carson shared a couple dozen pictures of my "new" family. There was Margaret the Opera singer and her daughter Martha, Carson's brother Ian, a fellow singer-songwriter, and his sister Carol, the clogger. There was Martha's sister Jane, whose Blue Note albums with Vicky Hamilton and Dave MacKay are on YouTube. And Carson shared music and pictures from his own career in the 1970s as the drummer for McKendree Spring, and many years of performing and recording with his wife Valerie on the West Coast.

In the mix of the pictures, somehow again appeared the common thread - pictures of our crazy Uncle Walter, the youngest of my grandfather's siblings, who somehow was the person who connected our family together in his alcoholic haze.

It's hard to wrap my head around how much it meant to find them. Their journeys were not easy, and they had to overcome a lot of challenges not of their own making. Family history can be painful for a lot of people. For each of us, this family connection brought great joy.

A day or so later I connected with Ian, the youngest of Martha's children. As I looked at his picture, I realized that I had met him once, at my grandparents' house when I was there visiting. I was probably 13 or 14, and my grandmother told me he was a cousin on my grandfather's side. That didn't mean anything to me. We just had fun goofing around for a couple hours. We then parted ways and I soon forgot much of anything about the encounter.

Suddenly here were Ian and I messaging one another, and sharing bits and pieces of a whole lot of lives lived in between that connection and this one. We both are indentured to the muse of musical creativity. To find that I am not the only singer/songwriter in the family was a revelation indeed. There is at least one other who hears these voices and feels compelled to bring them out of the ether and into the ears.

After we finished, Ian posted a video of his mother's last performance, at the Popejoy Theater in Albuquerque, New Mexico, playing in Sondheim's "Les Follies" in 2002. Martha would soon have a disabling stroke that robbed her of voice and career. Sadly, Martha passed away a few months before Carson and I sat connected to each other's ears across our vast continent. But here she was very much alive on my screen.

As the video came into focus, there was Margaret the Opera Singer's daughter Martha in her elegant glory, my dad's first cousin, with that platinum operatic voice pouring out of my speakers. My family, appearing quite real as they emerged from the web of mystery, right before my teary eyes.

> ***YOUR TREASURE: This happened because of building my tree forward as well as back in time. I use a strategy I call "successive umbrellas." First, identify your grandparents' siblings, their children and so on. The grandchildren of your grandparents' siblings are your 2nd cousins. Once you've exhausted that research, move on to your great-grandparents' siblings. What you know about their living descendants will be invaluable in evaluating your DNA matches, as invariably some of them will turn up – eventually! Remember to protect the online privacy of any living relative - obscure names and email addresses in any posts you share publicly.***

CHAPTER 12
The Last Great Nuclear Family Vacation

My parents decided to take their two kids to Ireland and Scotland to see from where we came.

In hindsight, it happened rather suddenly. Some of my discoveries and a few surprises in our family DNA tests somehow that led to the revelation that my parents had "always wanted" to go to Ireland and Scotland. Once they had a tangible excuse and some knowledge to spark greater curiosity, they initiated the idea of a family trip - just the four of us - no spouses, no kids. A "Last Great Nuclear Family Adventure," because it certainly wasn't going to be simply a relaxing vacation. Since we'd never gone on an epic trip like that, primarily because of the 10-year age difference between my sister and I, it was also the First.

The notion of going to Ireland was unreal to me. Of course, I always wanted to go, but such an excursion was a pipedream on an artist's lifestyle and income. I always say that I am rich in a great many ways, but those ways don't usually lead to sacks of unused cash in my closet waiting for an odyssey. Of the four of us, only Aly with her then-boyfriend (now husband) had been to Ireland, and that was a one-week grad school budget hostel jaunt spent biking around County Kerry, mostly in the rain. While our parents are about as far as one can be on the opposite side of the room from luxury tourists, they weren't going to have any trips like that.

In fact, when we agreed to the trip, Aly and I had committed to driving our parents around two foreign countries in a small minivan, on the wrong side of one-lane roads, for 16 days. Dad helpfully provided a lengthy list of things they wanted to do and see (such as stay each night with locals who have a piano in their house with which to unwind at the end of the day) and a more lengthy list of things to avoid (i.e. cities, and food of any sort that might include organ meats). I was ready for the trail of my ancestors. Dad was in search of a good steak. Who knew what kind of trip we'd have?

Paul was one of my first likely McKnight relatives I encountered at Ancestry. Over the months of carefully comparing our notes and documentation, we suspected that Paul's great-grandmother and Dad's great-grandfather were siblings, and their parents George and Ann McNeight would then be our common ancestors. His DNA results confirmed that we were indeed connected and that a shared pair of great-great (2G) grandparents were among the most likely culprits. Paul also happened to live in Newtownards in County Down, the very town where George and Ann raised their children between 1830 and 1850.

Once my parents decided that they wanted to go to Ireland, it was natural to make time to see this place and meet our Irish family. However, Northern Ireland did give us all pause. I'd grown up seeing the TV news clips about the car bombings and violence between the IRA and Unionists, and in particular I remembered the US news coverage of the Irish nationalist Bobby Sands and his hunger strike. I didn't understand the issues between factions then, (and I barely do now). Of course, with both the UK and Ireland in the EU, the 1997 Good Friday agreement had long ago removed the physical barrier that served as a flashpoint for decades. But it seemed there would be a hint of risk to go to the north of Ireland and particularly Belfast and Derry. So with mostly excitement and some lingering trepidation, we planned a day visiting our cousin in the hometown of our ancestors.

Dad is blessed with a few peculiar traits. One is that within a few days of leaving home he is physically overcome with the desire to return home. It's a powerful urge, to be sure. When I was seven, we went cross-country in our VW minibus. We traveled through and mostly around the big cities of the eastern US by night to avoid traffic, and hit a couple of national parks along the way before we finally saw the sun setting over the Pacific. Within a few days we were home, completing our 7,000 mile trip in less than three weeks, and that included the three nights in an Idaho hotel when the VW broke down.

Back when my sister was in college, he drove solo out to Nevada to pick her up from a seasonal internship. He had just retired and was excited to have an adventure. But once they connected and the van was pointed east again, the two of them drove nearly nonstop back to Connecticut in less than three days.

I don't know from whom he inherited his homing instinct. It certainly couldn't have been Gram, who loved to travel and would leaving Gramp at home to fend for himself, while she'd "gallivant" to various places with her girlfriends. Whatever the source of that gene, I was now going to be traveling with him if and when that mood struck. Barring a medical emergency, going home now was simply not going to be possible, and I didn't have a plan for that.

Dad is also a planner, down to the minute details. He might be politely described as a Type A++ personality. And so he wanted us to nail down the details many months in advance, which we did to the best of our ability. But we also knew we needed to leave ourselves some latitude to seize the moment. It's also not wise to count on two weeks of sunny weather in the British Isles at any time of year. Nonetheless, we had our plane tickets and rental vans booked in September for a trip late the following May, so we could focus on the what and where of the journey.... for the next several months.

The where was a week in each country. I was in charge of route planning, with the dual and occasionally clashing objectives of satisfying their criteria while hopefully visiting some places of significance in each of their family histories. My sister's job was to line up all the AirBnB accommodations, and then drive us around the countryside and backwards roundabouts for two weeks while her big brother navigated. It was certainly not going to be one of my more typical performing tours where most everything comes together on the proverbial spur-of-the-moment. Since Dad was more excited about Scotland, we decided to go to Ireland first, in hopes that the anticipation of Scotland would get our homing pigeon through the first week of our trip without getting edgy about going home.

Five days before we were to depart, we got an email from a new DNA connection. Our new cousin Eileen lived in County Antrim, and we quickly deduced she came from a third child of our George and Ann McKnight 2G grandparents. She was keenly excited to learn that we were flying to Ireland at the end of the weekend, and that Newtownards was already on our itinerary. Eileen quickly also showed her generous streak of Irish hospitality - offering to meet us all and arrange visits to sites of family interest. While she had never met Paul either, the two of them soon teamed up to plan our joint family excursion back into the past and making dinner arrangements for the near future.

I got up at 4 a.m. for the 400 mile drive to Connecticut to meet the others at the airport for our 6 p.m. overnight flight to Dublin. While I've been away on tour for two-and-a-half weeks several times, I had never had an ocean between my family and me, and thus my internal homing pigeon derived some irrational comfort that I could drive home anytime the need arose. The 10-year old was not at all happy that I was leaving. Somehow she got herself up when it was time for me to go, and her sleepy face resembled Cindy Lou Who from "How the Grinch Stole Christmas" when she confronted me in the bathroom for an endless hug.

I tried not to think about how far I would be from home for the next 16 days, or the undoubtedly uncomfortable and unsuccessful attempt at sleeping on a plane that awaited me. We would essentially be up around the clock for a full day and a half, and that would include our debut navigating across an unfamiliar country while driving on the wrong side of the road. We needed to get our minivan and provisions, traverse the southern half of Ireland on not much sleep, and somehow find our way to the right driveway and doorstep where we'd be welcome to come in and collapse.

But an unanticipated bonus of traveling around the summer solstice was long days! It started early in the last hour of flight, and it was already getting light as I wrote in my journal:

> "May 25, 2017. I am on a plane crossing the Atlantic, soon to touch down in the first rays of the rising sun in Dublin. It was exactly a year ago when cousin Lee messaged me that she was pretty sure she had found one of Margaret's family. Connecting with these "new" cousins and the many more discoveries that have ensued helped get my Dad excited about exploring his McKnight family heritage last summer, culminating in this trip to Ireland and Scotland upon which we are about to begin. It is hard to imagine all of the amazing things that have happened in this last year that brought us here. I am certain that I can't imagine the experiences and discoveries that lie ahead of us in the lands of our ancestors."

We had big plans for Ireland that would climax with our Newtownards visit and a final drive down the coast to Dublin. Our big plans didn't account for the big emotional heft that would come along with it. But in hindsight, how could it? This was the land of my people

- my mother's people. There were endless miles of scenery, shifting between the pastoral and the spectacular on the long ride to the western mountains of County Cork. We had no luck finding the two small villages where our forebearers lived before they made their break with the Old World. But the imagination flickered through the irrational notions, that every person we saw might potentially be a distant relative - a descendant of the Left Behind.

Visiting the seaport museum at Cobh, the departure point for more than a million Irish escaping the Great Hunger, and the last port of call of the Titanic, was powerful and potent. We found our ancestor Charles Kirkland on the list of the lost at the Titanic Museum. The images of people fleeing the famine, the quarters and conditions they endured on the seven to twelve week voyage across the Atlantic, as well as those Irish arriving in turn-of-the-century Ellis Island, seared the realities of their journeys into my consciousness. While the museum food court was bustling with happy and hungry tourists, our table looked out on the harbor, where our ancestors among millions of other desperate Irish slowly disappeared forever beyond the western horizon.

After our visit to Cobh, my mind frequently returned to the images and artifacts of those who had pulled up from everything they'd ever known. Their desperation did not obscure the fortitude and hope that they made the difficult voyage across the vast ocean with only what they could carry, and somehow withstand typhus and dysentery to actually set foot on dry land again. When they did, they started over from scratch in a foreign land, with plenty of barriers to even the most basic human dignity.

On our first night in the town of Donegal, Dad and I took a walk down into town to reconnoiter eateries and pubs that might be hosting a traditional Irish session or other live music. We wound up at twilight down by the water, and as I looked out over the harbor I realized that another family of Mom's ancestors had departed for Canada in 1826 from this very place. It seemed that their stories blanketed this land as much as the green rocky fields and flocks of sheep, as the familiar tunes of the past reverberated in the street.

The cliffs at Slieve League rise higher than Ireland's famous Cliffs of Moher, but there is less tourist traffic. From the lofty heights one can see a vast expanse of ocean that apparently has been well-traveled by humans for millennia. There were two options to get to the Cliffs - walk the three quarter mile roadway, or brave the hordes of walkers and drive up. Aly graciously drove the parents up to save their energy, while I opted to walk along the grassy clifftops and peer down at the pounding surf.

At an informational pullout on the ride up I'd been eavesdropping on an elderly guide telling his guests about the Neolithic civilization and the ruins that we'd visit later in the day. Whether or not his name actually was Paddy as he'd said, he was a delightful soul, grown sprightly and sage in part by a self-confessed healthy diet of pints of Guinness. He shared lots of interesting bits about his 85 years of life in County Donegal. And here at Slieve League, the octogenarian had opted for the walk like I had, while the others drove to the top.

I couldn't resist striking up a conversation, and our walk seemed to pass in moments. Before we parted ways, we both gazed out over the immensity of the Atlantic. I thanked him for his generosity in sharing his knowledge, and asked him, "What do you think of when you look out there?" Thinking that he might offer some lament for the four million Irish who left home during the Great Hunger, his response surprised me. "Vikings. I always imagine seeing their ships moving across the water, and of course I imagine that they weren't welcome. But it would have been fascinating to see them."

You learn stuff about people when you travel in close proximity for any length of time. I for one, never knew that my sister has a near obsession with ancient civilizations, particularly in the neolithic times. I didn't have much foresight that my camera would fill with shots of Mom and Aly crawling in and out of rocky crypts every chance they got, or that Mom and Dad, after nearly 60 years married by that point, still held hands when they were sharing in moments of amazement. Aly and I were lucky enough to catch them lost in the magnificence of a vista several times, serving as witnesses and discreet photographers too.

Our week had already been full of poignant moments, stunning scenery and an unexpectedly pleasant and physically easy trip. But none of it prepared us for the adventure that was about to unfold in Newtownards.

> ***YOUR TREASURE: If you take a DNA test, you'll want to reach out to your matches who may share your ancestry. It's best to be personal and friendly, and keep it simple and short, but give some clear synopsis of what you know that might be of interest to them. Keep records of who you've emailed, and especially nurture the connections with those who respond positively to your outreach. Not everyone will respond, not everyone who does will be enthusiastic, and that's how it goes. But that helps you appreciate the helpful and engaged cousins even more!***

CHAPTER 13
The Gift

Our first encounter with living relatives across The Pond is one we'll never forget!

When the big day came for us to meet our family, we were all excited and a goodly bit road-burned too. We'd been in Ireland for a week already, driving first the width and then nearly the length of the island, while exploring the coastline and the highest mountain roads we could find. Today was different though, for today we would cross the border between the Republic of Ireland into the United Kingdom - from kilometers to miles, and from euros to pounds. We set out from Donegal to the rhythm of the windshield wipers, enjoying the Irish mist and blooming rhododendron. With none of us being "city people," Belfast's bustling traffic set us on edge for a bit, but we made Newtownards in good order, and just as the rain let up we found the restaurant with minimal struggle. And sure enough, cousin Paul was there to greet us, with Eileen soon to join us after finishing her morning work.

Over lunch we shared pictures and stories, while navigating the considerable differences between our dialects and idioms. The sound of Northern Ireland, or "Norn Iron" as it sounds to outsiders, is quite different than what we had been hearing for the past week. I think Paul was amused when I returned unsuccessful from trying to settle the check. He told me "it's sorted." At first I thought that he meant the bill was sordid, as in excessively high. But of course he meant that he'd already picked up the tab and it was all sorted out, as they say. The dialect of Ulster will fascinate me to my dying day, but I might not ever readily pick up on it in conversation!

When we left, the town was bathed in glorious sunshine, and we walked around its bustling center. Eileen showed us the official state church in which our ancestors George McNeight and Ann McClements likely married around 1830, and the church where they would have worshipped and baptized their children, and the streets they would have walked some 175 years ago. In the town's central square the neat modern Edinburgh Woollen Mill outlet store sits in the space that not long ago housed McKnights Mens and Ladies Apparel. I don't know if these McKnights were connected to our family, but I imagine that ours struggled to keep clothes on their backs.

Our last stop in town was the 13th century Movilla Cemetery, where with some walking we finally found a single stone said simply, "Erected by Geo. McNeight." There was no record of anyone ever being buried there. Just another of the millions of Irish mysteries, except that this one belonged to us.

Eileen had anticipated all our needs with skill and grace, calling an end to our afternoon explorations to allow us time for a nap and a shower before reuniting for dinner. By dinner her husband, the Reverend Mark, and her brother John had joined us. The restaurant was on the waterfront in a coastal town near Bangor, the town where Dad's great-grandparents Andrew and Sarah had married and started their family before emigrating to Scotland. My seat at our table afforded me the view towards the southwest Scottish coast across that very same Irish Channel.

As we ate our meal in between the immense joy of getting to know one another, I sat mostly quiet for a bit, embracing and absorbing the gravity of the moment. To the best of my knowledge, Dad's great-grandparents and their family left here and never returned. Our family union was likely the first time members of these three siblings' families had been in the same place together in over 150 years.

After dinner Eileen mentioned how they would love to hear me play. When I travel, a guitar is usually as essential as a toothbrush and clean clothes. The logistics of our airplanes and travel precluded me bringing one, and I'd been missing it despite spending most of our waking hours driving, eating or walking someplace. We were also several days removed from Dad's most recent "wind-down" piano. But Eileen had foreseen this omission as easily as every other detail; Mark had brought his acoustic guitar.

I'll never know if Eileen knew that there was a grand piano in the foyer when she made the restaurant reservation. It wouldn't surprise me. It would surprise me if she knew that it had three stuck keys. But with a chance to make music together for a little bit, Dad wasn't going to waste the opportunity to play or to bring home a story about "making do" on that piano to anyone who'd listen. After all, we were in a land of stories and it wouldn't do to return without some of his own. Since I had left my guitar picks with my luggage at our bed and breakfast, his story would gleefully include my use of a folded-up business card as an emergency plectrum.

So there we sat, as patrons walked in and out smiling at the music of an American father and son. We played some of my songs and a few of the jazzy classics from the Great American Songbook. While we had been largely steering clear of the world news, we heard word earlier in the day of the passing of Gregg Allman of the

Allman Brothers. As they had been a big influence on my musical upbringing, and were a favorite of Dad's too, we wrapped up our lobby session with "Stormy Monday," an old blues tunes the Allmans had made famous.

As I put the guitar back in its case, I thanked Eileen and Mark profusely for being so thoughtful, as I had certainly been missing my guitar. With a twinkle in her eye, Eileen said in her lovely lilt, "It goes with you." I was sure I misunderstood her, as we'd been struggling a bit with the accents and idioms since we'd crossed the border that morning. But Eileen and Mark were steadfast, and they assured me it was their intent I take the guitar.

A beautiful Lowden steel-string guitar, made there in Northern Ireland - Mark had owned it for some 20 years but played it rarely. Now I stood before my new cousins completely dumbfounded, still holding the case. She finished me off with, "We want our American singer/ songwriter cousin to bring home something of his homeland."

As the incredible reality sunk in, we all realized that something incredibly special and precious was happening. Despite the late hour, it was still plenty light outside and we were all reluctant to part ways. We took some photos together in the twilight by the water - the very same water that had separated our families 150 years ago, connected to the vast ocean that would soon separate us again.

When we got back to our cottage that night, I went to the Lowden Guitars website to see what more I could learn of the incredible gift. I was able to look up the serial number and found it was built in 1994. It took a moment for the final fact to sink in. My gift guitar had been made in the factory George Lowden built in Newtownards, the hometown of my ancestors.

We still haven't figured out who, if anyone, is buried under that gravestone at Movilla.

> ***YOUR TREASURE: What tangible items do you have from your family history? An unknown treasure or some of their letters? Some people spend their lives accumulating stuff, and going through their effects when they are gone can be daunting. I try not to spend much time lamenting what was or may have been lost. But I do recommend taking care to save those "old things," lest we come across an explanation someday.***

CHAPTER 14
Exchanging Soil

Following our ghosts to a grave in Scotland, while my dad once again surprises me with gestures of sentimentality.

Once our family made the decision to go to Ireland and Scotland, it was embedded in the planning that we would see as many known family locations as possible. Given that our itinerary included both the Isle of Skye and Orkney Islands, as well as the mountainous Scottish highlands, we had some inkling that we might overdose on spectacular scenery as well as long drives. And so a family history adventure in the more pastoral lowlands below Glasgow seemed like a good change of pace before we returned home.

I'm not sure exactly when or why my family left Newtownards for Scotland long ago. While the potato blight and Great Hunger of 1845-52 affected Northern

Ireland, it wasn't as severe as it was elsewhere on the island. Clearly things must have been in some downturn, for in the 1850s and 60s the Ulster Scots crossed the Irish channel in droves to work in the booming greenstone and iron ore mines of the Scottish lowlands. Dad's great-grandparents Andrew McKnight - aka Andrew the Miner - and his wife Sarah Milliken followed her parents from Northern Ireland to Dalry, in Ayrshire province. Eventually they moved on to West Calder, a small town 20 miles southwest of Edinburgh that boomed with the oil shale industry late in the 19th century.

All my reading and research suggests that my family had a hard go of it. Andrew and Sarah lost babies, small children, and a 20-year-old son who burned to death in a naphtha fire. Loss appeared to be a near constant companion to hardship. The paycheck from the mines may have been steady, but it was nasty work, unhealthy at best and deadly at worst. And despite the fact they were of Scottish ancestry, they were not welcomed as such by the native Scots. Rather, they were treated with disdain and prejudice because of their Irish origins.

Somehow Dad's grandfather Andrew the Fiddler earned his way into the University of Glasgow to pursue a teaching degree - a highly coveted ticket out of the mines. Eventually he and two of his adult siblings would make their way to America.

The other siblings remained with their parents in West Calder and the surrounding area. I have no idea how long they stayed in contact with their American siblings. But any regular contact across the sea faded away many decades ago. By the time I started climbing around my family tree, the various lines of descent had drifted away from any real knowledge of each other.

While I knew the towns where our family had lived and worked, I really had no idea what might remain to be seen, or if anything we did find would have been of any significance to them anyway. The mines closed long ago, and the towns are close enough to the cities and train lines to have become essentially suburban communities in our century.

The Scots, like their English neighbors to the south, are diligent about recording facts. With a bit of help from some of our cousins, we found vital records for many of the family, as well as some of their descendants - including a handful that were among our solid DNA matches. I knew that Sarah had died in West Calder in 1901, and Andrew the Miner died in 1915. I was pretty sure they were buried there someplace.

A couple of years after starting this family history research, our trail led me to Kerrie, a 3rd cousin in Scotland who also descended from Andrew and Sarah. I introduced myself across the wires to share what I had learned thus far. Like so many of my fellow family storykeepers, she'd largely been working alone and didn't have a lot of close family with whom to connect. When she heard that we were planning a trip, we tossed about the idea of meeting - and hoped we could still find a place of some family significance.

A month before our trip, I posted my theory about Andrew and Sarah being buried in West Calder to an appropriate local internet forum. And much to my surprise, I learned that a local family history group had transcribed the burial registers for all the provincial cemeteries. Before long we knew where they were buried together with their son John and a child of their son William, and that they were in an unmarked grave, but adjacent to a large headstone.

Now we had a destination. But in my living room 3,000 miles away, finding it felt like a tall order. For Kerrie however, this was exciting new information about her ancestors, and close to her home near Edinburgh too. She made an advance reconnaissance visit to the cemetery and found the graves. Now all we had to do was make it across the ocean, survive a week in Ireland, fly to Scotland on a small plane, make it to West Calder and meet her somehow. The rest would just have to fall into place.

As a rule, my dad is not a sentimental guy. So he caught me completely by surprise when he expressed a desire to bring a little soil from his grandfather's gravesite in Forestville (Bristol) to return to the "Auld Sod" of Scotland. As it turned out, I had a concert in Bristol just two weeks before our trip, and a few blocks away from the Forestville Cemetery. On a rainy spring evening evocative of Scotland, I made a quick pre-show pilgrimage to visit Andrew the Fiddler's grave and fill Dad's two small film containers with dirt.

While Ireland alternated between shades of pastoral to spectacular, Scotland was nonstop in-your-face incredible....assuming the clouds lifted enough to see anything. We had the luxury of staying with the family of one of Aly's ornithology research friends near Loch Ness, and using their bed and breakfast as a base for our divergent destinations across the Highlands over the week. The day before our Scots family history mission, we would head south past the majestic massif of Ben Nevis, the UK's tallest mountain, and through the spectacular rugged mountains and valleys of Glencoe. We'd been drowned in rain the day before, but this day only dawned drizzly and grey. We lingered a bit over breakfast, thinking after coming all this way that our ultimate objective in highland scenery would evade us.

As luck would have it, we dawdled long enough to see the clouds slowly lifting as we made our way to Fort William. Just as we made the turn into perhaps the most stunning scenery in the Scottish Highlands, we were in brilliant sun and blue sky. At the first roadside pullout, the parents stood arms around each other staring out in wonder and amazement at the craggy ridges rising nearly straight up on all sides. I said a hundred silent thank yous to the universe for pulling back the veil of clouds and allowing them these moments.

We squeezed every ounce of light out of that postcard-perfect day. We took detours and side roads, stopping every few minutes to gawk at waterfalls spilling off every high cliff, and at treeless meadows. Of the many gigabytes of video and photos we took on our trip, the mountains and moors of central Scotland demanded the biggest allocation. We devoted a few of those bytes to capturing our parents, standing hand in hand relishing their incredible adventure.

In all of my life of being around Dad's music, I had never heard him sing. As we drove along the sunset-lit waters of Loch Lomond towards our AirBnB, an unfamiliar Irish-sounding tenor began warbling "Wee Bonnie Lasses" from the back seat. My sister quickly joined in, this being both a tune and a ritual apparently quite familiar to the two of them. They carried on that way for a while, while I sat in mute astonishment. I was too stunned to turn on my phone's voice recorder, but somehow it felt like the moment would have vanished if I had. It was my role to bear witness and let it play out, and no more.

Our final day dawned sunless, a dreary return to more suitably miserable Scottish mist interspersed with outright rain. We made our way down busy roads along the heavy industry of the Ayrshire coast, looking west at the many islands, as well as southwest towards the invisible coast of County Down. The rain gradually morphed into a mist more suitable for chasing ghosts. We rolled into West Calder past several enormous piles of 19th century mine tailings that had transitioned into a 21st century park.

We found a parking spot a block from the café, all of us excited to meet another relative. My sister was now a seasoned veteran of all things vehicular being backwards, so navigating narrow streets and tight parking had become old hat. As for me, even on Day 16, I was still triple checking myself when crossing streets. It made me a steadfast proper pedestrian, lest my American jaywalking ways end my life close enough to be buried with my great-great-grandparents. I didn't come 3,000 miles to get flattened in their very town, and certainly not without meeting my cousin.

We had exchanged pictures with Kerrie before the trip so we would easily recognize each other, but we needn't have worried as we had the place to ourselves. Her resemblance to Aly was uncanny, despite the distance of our cousinship. Over our leisurely lunch she shared pictures and stories about her family, descendants of my Andrew the Fiddler's older sister.

After a delightfully lengthy getting-to-know-you session, we went to complete our mission and pay our respects to our ancestors Sarah and Andrew the Miner. Kerrie brought a beautiful flower bouquet. Dad carefully emptied his vials of soil, and planted a couple of small American flags he had brought so visitors would know that those at rest had received family from the States. Once again, he surprised me when he collected a little dirt to bring back to their son's grave in Forestville. Between the sudden sentimental streak and the outburst of singing, I was beginning to wonder if this whole crazy experience had changed him a bit too.

It was an exhausting day, and the thick cloak of grey Scottish clouds hadn't helped. After parting ways with Kerrie, our passengers marveled for awhile about how much she had looked and felt like family, before they fell into a deep sleep for the rest of the drive to Edinburgh.

The mountainous old city is a busy rabbit warren chock-full of cars, people and noise, and navigating it was as stressful a driving adventure as we'd had anywhere. It took a while to wind our way to our AirBnB. But the glorious view of the majestic ridge known as Arthur's Seat dominating the skyline in Holyrood Park from the deck in the late daylight was a pleasant reward.

By this point, my family was done. Not just for the day. They had essentially reached the finish line after 16 days and a nearly 10,000 mile voyage, improbably and dare I say, almost gracefully. They were spent. Cashed out. The need for food got Mom and Aly out for a 10-minute walk to a pub. We were still home and in bed before the sun dropped completely out of sight. In the morning there would be airports, customs and security, planes and long hours of sitting to be endured.

As for me, I found sleep elusive and restless. After all, we had just finished the last of what we set out to do. We'd followed our trail, from the town where our ancestors had lived a good portion of their lives, to the mining villages in the lowlands of Scotland, and over an ocean home to America. We had visited ancestral graves, hugged cousins alive and in the flesh, and walked in village streets where chapters of our stories had long ago been written. It was a lot to process at the end of a long journey that included months of planning and years of research. While my flock of homing pigeons and I were ready to return to our respective roosts on a far shore, the first and last great nuclear family vacation had also successfully avoided fission. I quietly played my new guitar for a few minutes, wondering how different the me was who was returning home.

After the long day's journey back to America, and the next day's drive back to my home in Virginia, I realized the answer was - a LOT! Back in my cozy kitchen, I poured a wee dram in honor of Andrew the Miner and his beloved Sarah, my great-great grandparents. They lost at least six of their children. Three more grew up and left for America. Even now I regularly connect with more of their descendants. Considering all of the generations going forward, I'd wager there are many hundreds of us.

I am certain that they didn't give thought to how much of their DNA would be roaming the globe, four or five generations into the future. I'd imagine that they would be astonished, and maybe a bit pleased and even proud too.

> ***YOUR TREASURE: Thanks to the free crowdsourcing websites FindaGrave.com and BillionGraves.com, you don't have to visit the graves of your ancestors to see the stones and learn valuable information about their lives. Thousands of volunteers continue to add pictures and details about cemeteries around the world.***

CHAPTER 15
My Morning Walk

Appreciating the gift of recording our lives and stories, while the rest of the world passes without our notice.

I begin my walk this summer morning at just past 8 a.m., with the temperature already well above 75°. A high-pressure heat dome over the mid-Atlantic brings us the first real gripping heat wave of the summer, with pea soup humidity holding everything still, even at this early hour. I am still flush with the discoveries and experiences of our recent trip through time. There is much yet to process.

My gravel road walk this morning is anything but lonesome. A cacophony of birds and insects raise their protestations against the sweltering humidity. A hawk swoops by right in front of me, likely in response to the audible cry of another nearby hawk. A cardinal whisks down into the road in front of me and skims away with its seedy treasure. A sidewinding fly zips past in a Doppler effect of whizbuzz approach and retreat. An orchestra of fiddling crickets saw their wings back-and-forth, providing the soundtrack to life in the nearby tangle of vines and locust trees.

Countless creatures will be born this day, and myriad others will perish. Some of those will expire at the end of a typical life expectancy, while others will pass suddenly in the clutches of some predator. That hunter's success this day will prolong its life and perhaps that of its offspring for some time, in the way of things. Essentially all these cosmic arrivals and departures will go unnoticed by the humans who have shaped and manicured this landscape.

Taken individually, these creatures' lives will have little impact on their greater ecosystem. But the ripple effect of their collective accumulated consequences gradually shifts the parameters over time. Alien species will arrive and take hold, perhaps significantly altering that ecosystem. Some insect empires will thrive and rise....until the hungry predators are drawn to gather for the feast, remaking the roles and players in the ever-continuing theater of life in the Crooked Run Valley. In their wild world, this anniversary each year will go unnoticed, much like most of the human anniversaries tethered to this position in our magnificent blue ball's orbit around our sun.

We humans are the only species known to record specific events to keep a historical record for future generations. Even so, only the biggest events touching large segments of our human ecosystem get included as part of our collective story, perhaps to be remembered in the future as the calendar flips by this date again. The earth continues its solar journey, oblivious to us and our doings.

Somewhere a stranger's spontaneous random act of kindness might save someone's life, that they may grow old and make their own mark on the world. Meanwhile somewhere else, someone uses their power to commit unspeakable sins against a child. Up until recent human history, everything other than the most notorious and noteworthy events would be preserved primarily in the memories, mementos and conversations of those who lived through them.

Being raised by others is far from a uniquely human experience. What again is unique to our species is most adoptions are recorded in some way, often now with the eventual full knowledge of some or all parties. Sometimes that knowledge is held from the adoptee until the secret slips out, which is now a frequent occurrence thanks to the advent of DNA testing.

My dad had a close DNA match turn up recently with no tree attached. The match amount seemed to be roughly in line with a 2nd or 3rd cousin relationship, which would indicate that this woman from Liverpool and Dad share a set of great or 2G grandparents. A match that close will always get my attention, especially if it's not fairly obvious where the connection is.

Kim responded enthusiastically when I sent her a message, as she is an adoptee eager to learn anything of her genetic origins. Within a couple of months and a couple of strategically chosen new DNA testers, we were able to string together the connection through the known shared matches that linked her birth mother with Dad's paternal family - a descendant of Andrew the Fiddler's older sister. Once she got her birth records unsealed, she now knows she is as connected to our Ulster Scots genetic inheritance as much as any of us.

While the entirety of our connection happened by launching electrons back and forth across the Atlantic, her messages clearly conveyed the gravity and poignancy of now knowing more about who she was. She has lived a full life including raising two adult sons who've also turned out to be fine guitarists. I could not viscerally understand how much this was a huge treasure for her to find family and to learn more of her own story, but I certainly could cherish and honor her feelings.

It doesn't always go like that. We don't know the personal circumstances of her origins and what heavy emotional turmoil may have surrounded them. It isn't hard to imagine a range of unpleasant situations resulting in a baby being given up for adoption. In many jurisdictions around the world, the records are sealed, and the adoptee is not allowed access to their own stories so long as the birth parents are alive.

Until now, I hadn't ruminated much about what it would be like to have my birth and origin stories essentially be a stubbornly blank canvas shrouded in mystery. I've been blessed with a rich inheritance of stories and lineage full of colorful flawed characters. It is a luxury that my interest is driven simply by curiosity about my family, and I'm happy to occasionally be of value to other researchers who have reached out to me about a shared connection. I hadn't considered the magnitude of spending years searching for one's roots, and then receiving such information in essentially a thunderbolt.

We are fortunate then to be neither cardinal nor cricket, for the basic details of most of our lives are recorded in some way or another. In theory, many of those details will be accessible to our descendants someday long after we are gone. We are unique among the world's species in our capacity to retain memories of specific people that last far beyond their lifespan, aided by our written language. For at least these last couple hundred years, the basic milestones of births, marriages and the inevitable death, are dutifully recorded in some official place a good percentage of the time.

Of course, until the advent of digital technology and social media, beyond those milestones the vast majority of human lives passed without record. Yet our ancestors long ago and far away sat down to eat, raise children and worry after them, secure food and work to prepare and preserve it, and endure great disruptions in their worlds due to war, weather or new tools and inventions. In their way and time, their lives were as rich and full of life as are our own. These details of all but the most recent generations will never be known to us. And even then, the only personal trace of many of my ancestors is the "X" they used to mark by their name on an official document for the birth - or death - of a child.

Those ubiquitous, wily and ever-evolving strands of double helices hold all of the stories, even though we've learned to decode but a minute fraction. We are in essence composed of story as much as those strands, and those too are ever-changing. Once in a while, it is a small segment on a single tiny chromosome that unlocks a large piece of our past, raising the veil on an unknown and unread chapter of how we got here.

In that context then, how lucky we are not to be the bee or the squirrel, but to have both the curiosity and the means to keep some of our stories. It is more possible than ever before to access information and piece together the puzzles of our predecessors. And perhaps that explains our sometimes surprisingly emotional reactions to finding tangible connections to the mundane details of an ancestor's life. A letter, a favored recipe, a photograph from a child's baptism - we correctly associate those things as part of us, and somehow part of our story. It is as though from a collection of dry dates and facts, a real person suddenly emerges.

It is another reason why being human is special indeed.

> ***YOUR TREASURE: If the unthinkable happened to your house, what would happen to that treasured picture of great-grandma, or that 3G grandfather in his Civil War uniform? Those precious old pictures and letters need to be scanned at a high resolution, and saved to a safe place on the Internet (consider a cloud service like Google Drive or Dropbox). Not only does this mitigate the disaster in event of a fire or flood, but it allows you to selectively share and collaborate with other living relatives. Even though technology changes at a dizzying pace, we can still read words on paper from centuries ago. So it will be too with our digital matrices of 1s and 0s.***

CHAPTER 16
Lincoln

Where I put down roots is different from the home of my father, as is the place where he settled when he left home. Living amidst history makes me more keenly aware of why it matters to know "where we've been," as well as the role of communities as keepers of heritage.

There are often sudden bolts of realization and clarity for those who study history, and family historians are no different. It struck me one day that my line of McKnights share a common thread as far back as I know our story. Starting with a young George McNeight in Ireland apparently leaving his family village near Annaghmore and raising his own family in Newtownards, each son in my line of descent was born in one place and left to live our lives someplace else. Whether it's a conscious decision to escape whatever bad mojo we associate with our birthplace, or simply moving for a better opportunity, each of the seven generations I've traced of my dad's family moved away to begin their own story.

And so in keeping the family tradition, I blew southwest from eastern Connecticut's hills as fast as the wind and the interstate would take me to a job in Loudoun County in northern Virginia. Once the northernmost edge of the Confederacy, Loudoun stretches westward from Dulles Airport to the crest of the Blue Ridge and Harpers Ferry, bordering Maryland across the Potomac River to the northeast and the Shenandoah Valley to the west.

When I arrived here, more than half of the county was still prime agricultural land. Most of it was already planned and destined to become new suburban cities that I simply couldn't imagine. The eastern and central thirds of Loudoun are now an outer suburb of our nation's capital, as well as home to a large portion of the servers vital to the global Internet. Sadly for much of the county, the rich soils of that past are buried under tens of thousands of picture-perfect lawns in cookie-cutter neighborhoods, and many square miles of new pavement.

Here in the western part of Loudoun, my still-rural village of Lincoln endures, within sight of the Blue Ridge and visibly steeped in American history. Around here, the past and the present swirl around each other in a fight for the future. Several of my neighbors live in houses built in the mid-1700s, and some of them live on land granted to their family in that same era. Compared to my McKnight ancestors leaving already ancient places like Newtownards and West Calder, a couple of centuries is not much history. Then again, when one can easily find fields of gleaming facades in brick-lined downtowns of Go-Gettersville sprouting where soybeans flourished a decade ago, a few old houses, farms and unpaved lanes feel nearly timeless. An information technology specialist can work an early winter day at a server farm in densely developed Ashburn, and then return home twelve miles west into the early evening darkness to find that a bear destroyed a bird feeder or got into the trash cans.

The fertile Piedmont collides with half-billion-year-old bedrock of these beautiful Blue Ridge hills, lined with pages from nearly every chapter of the American story. The story of Lincoln is a complex stew, some of its ingredients buried deep in the layers of red clay, while others lie close to the surface like the rocky outcrops in most any local creek valley.

The echoes of the past are readily heard in our little village, home to the longest running Quaker meeting in the Commonwealth. The Quakers who settled here in the 1740s were part of a large migration from Pennsylvania down the Shenandoah Valley, along what was then the frontier. Several currently occupied homes were built in those early days.

Originally known as Goose Creek, Lincoln is in the heart of "Mosby's Confederacy," the local moniker describing the territory of the notorious and nefarious Confederate guerrilla raider. In spite of the community's many abolitionists and Union sympathizers, in late 1864 blue-uniformed troopers brought devastation by torching barns and granaries to quell any local support for Mosby and his men. After the war the abolitionists renamed the village in honor of the martyred 16th president.

We live on a dead-end gravel road, and yet we are a three-minute walk to the Post Office and center of the village. For six years I walked my daughter to our many decades-old school at the other end of Lincoln. The original one-room school built in 1814 still sits open to visitors. My neighbor Alan Cochran was raised here, and he says that the view from the crossroads at our end of the village hasn't changed since he was a boy 50 years ago. He's probably right.

Our humble old house is built atop a stone foundation and an unfinished basement, which in many ways reminds me of Gram and Gramp's house. We bought it in 2009, at the bottom of a historically bad housing market and near the bottom of the interest rate curve. Most of our neighbors are pretty quiet. The Quaker Burying Ground is across the little valley from our front yard, while the cemeteries of two historic African-American churches border our backyard. The dead probably outnumber the living here by a good 5-to-1.

Our house's previous owners Maud and Kenneth Smith are among them. They raised their family here including several foster children. They attended the historic African-American Grace Episcopal Church, and are buried in the cemetery which abuts the very back of our property. Like many of their past and current neighbors, for much of their lives the Smiths surely endured many of the cruelties of segregation and white supremacy.

I am lucky to live where rural beauty is still seen as an economic asset. While the landscape around our village is still largely pastoral, I've seen enough change in my quarter-century to understand that our history provides but a temporary bulwark against the inevitable spreading oil slick of suburban sprawl. Since 1990, Loudoun's population swelled from 90,000 to more than 400,000. Over these last three decades, Loudoun has been the fastest growing county in the nation several times, and by most any measure it is one of the wealthiest counties in the nation. Two to four gleaming new schools go up in different parts of the county each year to accommodate the flood of newcomers. The eastward view from the top of the Blue Ridge looks a lot different than it did to Mosby's Rangers scouting for Union troop movements back in 1864.

The cultural changes I've seen here have been dizzying too. Where decades ago Loudoun's culture was largely defined by segregation and the tension between its white and black residents, a cosmopolitan mix of literally hundreds of languages from dozens of countries now bursts at the seams. One can find pho, naan and falafel made fresh on many of those shiny blocks of new downtowns.

It is also a place of incredible scenic beauty. The Blue Ridge forms our western border, with the Appalachian Trail along its crest, nearly midway through its nearly 2200-mile trace from Georgia to Maine. The rolling hills are home to dozens of vineyards, specialty farms and other rural/outdoor tourism enterprises. Just a decade ago we still had over 300 miles of gravel roads, which the old-timers might say is payback for the unreliable loyalties of this border county during the Civil War. That network of gravel roads still provides the main access to many of those destinations.

Stories depend on remembrance, and remembrance requires both memory (written or oral) and accessibility. We cannot possibly keep all the stories of the world, but we can certainly lose much of what we've kept. Sadly I've seen many of the world's irreplaceable and precious cultural reservoirs of story lost to war or disaster, even as new technology enables historians, archeologists and anthropologists to make exciting advances in our knowledge of the past.

Understanding a place over a continuum of time helps to better contextualize the stories, choices and actions of the people of that past. It is crucial to understand their basics of life - how they lived, how they traveled and the work they did - as well as the impact of advances like home electrification, indoor plumbing and the automobile that radically changed everyday life.

Loudoun County is locked in an ongoing struggle between a relentless onslaught of development and the preservation of vital local heritage, including the final resting places of enslaved and indigenous people. We are rushing to save and understand our more complete story before it is forever lost to the work of the landscapers and the builders.

I'd like to think that the time I've invested in my own family's stories in some small way honors the people who have lived here in western Loudoun too. Other people's ancestors lived their stories out in these hills of our Crooked Run Valley. They traced lines to other homes, work and communities, much as my ancestors did in places far from here. Perhaps by sharing my own journey, I might help others to see this land has worth far beyond its value as real estate.

Conversely, I hope that the current residents of my ancestral places like Newtownards and West Calder, when confronting inevitable conflicts between past and progress, continue finding some inherent value in preserving it in some fashion. Maybe it is easier to see the worth of places that have existed for many centuries

- communities whose identity totem includes that history.

A friend of mine speaks of being "exiled from our childhood," a notion I find powerful and poignant. From my vantage point here in middle age, it seems each generation goes through a series of awareness events - where we realize something we knew has transformed or disappeared. That awareness begins as early as our tween years, when we find a treasured toy or have a sudden memory of something from our early childhood, and know we have outgrown it. It may be apparent that it is an experience we will never have again, and often these revelations are among our longest-held stories about ourselves.

As we raise our children, we are keenly aware of how so many things are different in their world. We witness the passing of familiar icons and/or experiences we treasured, like the carhops at the A&W Root Beer Stand or going to the drive-in movie theater. We may explain it to our kids, while realizing that they won't fully understand until they have their own experiences.

To paraphrase Thomas Wolfe, we can never truly go home again, and we can never return to our childhood either. Perhaps for some of us, reclaiming small elements of our cultural heritage from select ancestors offers us a bit of balance or relief from that inevitable loss.

My little village has taught me much about keeping our stories safe and accessible by continuing to tell them. History can make us feel infinitesimally small and inconsequential, and in the same breath, so special and unique. Perhaps we have some duty to ourselves and our communities to play a small role in our story keeping, or at least to do it no harm. Maybe we are lucky enough to recognize the deeper significance of a moment we know is fleeting, and to have the luxury of being fully present as it happens. It becomes a part of our story, and in some way we are a part of it; "I was there when...."

> ***YOUR TREASURE: The place you call home has stories of interest to others too, in its cemeteries and records. Can you help those looking to learn about their ancestors who lived where you do? Consider joining a local history group online or in person. You may be fascinated by the stories of your home, as well as occasionally being an "information angel" for someone who lives far away.***

CHAPTER 17
How My Great-Grandparents Raised Me and My Kid

Puzzling over my grandfather's family led to some startling insights and epiphanies about the legacy of our parenting.

The world of psychology is full of courses, books and experts expounding on the need to understand childhood in order to understand fully formed personalities. There are also scores of learned people defining and debating a wide range of perspectives on the never-ending nature versus nurture argument.

As I've aged, I've spent a lot of time pondering the impact of my parents' childhood experiences on my own life. Since I grew up with both my biological parents, I assume that whether by nature or nurture, I inherited and was impressed upon by their traits, much of which were developed in their childhood. But I find it likely that would happen with whomever raised us - that the genetic connection to our ancestors doesn't affect us as much as those who serve as our parents, and how they were parented.

When I was an undergrad at Connecticut College, I took a basic American History course at night one semester to fulfill a core requirement. Dr. Michael Burlingame, a scholar of Abraham Lincoln with a passion for psychohistory, held me in near complete thrall for our three-hour class each week. An exuberant personality with a large sense of humor, his approach to understanding key figures through the lens of how their personalities were formed in childhood was fascinating. I wound up taking two more classes with him, largely because I had never even thought to view historical figures in that context. It was like I had been given a Rosetta stone to dive into the fascinating personal stories of famous people. I had no idea that decades later I'd be using that tool on my own family history.

I have heard plenty of stories from my parents about their childhood. It is only now in my middle years that I really am beginning to understand the arc of our story. One might think of the lucky members of my generation as a bridge between aging parents and raising children. I have the luxury of my parents still being available and lucid to answer questions and share experiences. I believe that my parents and I each give context to my childhood experiences, which I simply couldn't appreciate earlier in life.

Mom lost her dad when she was 12, the same age my daughter is at this writing. She didn't know him as an adult, and as the youngest of three, didn't know him in the same way that my only child knows me. Her parents' absence in our world has been a vessel for my whole life in many ways. My daughter has constantly been a lens for how I see that vessel - if my time with her ended now, what would she remember? I have a more than fuzzy notion about the magnitude of that loss.

Since Mom's mother died right after my parents were married, Gram became her main role model as a mother-in-law and grandmother. She accepted her as family with open arms, encouraging words and a loving heart, while always lending a hand and goading the rest of us to do more around the house because, as she would say, "Poor Kathy works so hard." Since my grandparents came to visit us in between holiday trips to their house in Bristol, I had a whole lot more direct childhood contact with Dad's family. And since my Gram's family was small, she had one sister with no children, that meant most of the family was McKnights.

Ahh, the McKnights, Gramp's people. While my grandmother was sweet, smart, cultured and hard-working, Gramps was impatient and could turn harsh in an instant. He was blessed with immense intellect, but could suffer no fools. If someone crossed him, they were out. Done and gone. Used up with no second chances. And apparently he and his siblings, growing up with their cold and stern immigrant parents in a small house, had enough of each other by the time they were adults. While he came from a family of eight children, three of his siblings are largely a mystery, even now with all our advanced research tools brought to bear on them. I've never even seen a picture of them.

Why most of his siblings went off the rails in one direction or another may never be completely known, but alcohol and general mental instability may have had a lot to do with it. Looking at the bigger picture, I'm fairly sure my great-grandparents just weren't good at parenting, plain and simple. And it would seem to follow that their kids largely careened through their lives as a result.

I met three of Gramps' siblings. He had no stomach whatsoever for his younger sister Isabel, but she was fond of me. "Dizzy Aunt Izzy" played the organ, painted, smoked a lot and lived in a trailer one town away from my grandparents. She would have loved to fill my ears with stories about her siblings, and she probably did. But since I didn't know any of them, those stories went in one of my ears and continued on through. I do remember her telling me she hadn't spoken with her only daughter for many years, and me being unable to process how that would work.

His oldest sister Sarah married into an old Massachusetts family, and lived a far different life away from the bustling cities of Bristol and Bridgeport, including many years on a dairy farm in far northern Vermont. I made the trip to meet her when I was a teenager, full of curiosity about my grandfather's sister. She was elderly at that point and living with her daughter's family in northeastern Massachusetts. I don't remember her saying a lot, and I remember not being surprised that she didn't. I knew too that she wasn't much for long-haired men, and at that time I was one. She and Gramps shared a streak of responsibility, and I've come to appreciate her now for diligently keeping many of her parents' effects after they had passed away. They've been invaluable to me in simultaneously better understanding my family and muddying the waters.

Gramp's youngest sibling Walter was an incorrigible drunk and....an expert truck driver. My cousins tell stories of him parallel parking his 18-wheeler in their cul-de-sac like he was parking a VW bug. Uncle Walter was always up to some scheme, usually needed a little cash, and was pretty consistently a solid three sheets to the wind.

The funny thing is that somehow he kept ties to family across the distance, even as he burned and blew up bridges with his own son. As my 2nd cousins and I began connecting with each other over these past few years, we found that despite knowing next to nothing of their McKnight family, they all had calls from Uncle Walter as kids, with three common threads. 1) He was always good and drunk when he called, 2) He always said "You know you're a Scot, don't you?" and 3) He wouldn't hesitate to drop a manipulative, "You don't want to talk to your old Uncle Walter anyway," which was usually true. They had no idea who he was or how they were related to him, only that he was this crazy voice on the phone long distance from the East coast.

What I knew was that Dad's childhood was complicated, and in big ways marked by his father's alcoholism. Gram was working as a nurse at the hospital, while Gramps was an engineer designing and building machines at New Departure Bearings division of General Motors, turning out ball bearings for the war effort. Eight years younger than his only brother Doug, Dad was pretty much a latchkey kid.

Gramps was in engineer mode 24-hours a day during the week, so once he met his buddies at the bar on the weekend, he didn't waste any time letting off that steam. And when the bar closed, he had no inhibitions about bringing his motley crew back to the house for a few more belts, at the expense of any pie my grandmother might have left out to cool for the next day's church event, and of his young son, whom he often woke up to show off to his fellow drunks.

Dad learned from a young age that weekends were a great place to be elsewhere, when his father would be reliably "well-oiled." Even after Gramps sobered up around the time I was born, the relationship between father and son was always strained, laced with a critical air toward one another.

I should say that for all my grandfather's mechanical brilliance and Scots temperament, I believe that I got to enjoy the best of him. I wear his name, as he wore his father's, and he wore his father's before him. Gramps was sober by the time I was born, and he liked having a grandson to take fishing and skiing. I came along in the middle of an 18-year gap between the other grandkids. He loved baseball too, so we had common interests.

But for his siblings, there was no love there, and no connection either. He went about his life with his friends and his nuclear family. He hated spending money (did I mention he was a Scot?) so he bought a moped to save on gas and he packed his own shells to take skeet shooting up at the gun club. Gram probably half-expected that at some point she'd come home to find he'd blown the house up, but she always gave him great credit for his smarts and common sense. He went skiing most of the winter right up until the winter before he died.

Gramps beat cancer three times, but the last time took most of his tongue and the roof of his mouth, and so for his last few years he couldn't articulate his words clearly. By the time I was old enough to have adult conversations with him, he was nearly impossible to understand, and too impatient to spend any time trying to be understood.

As I gradually pieced together bits from the many McMysteries about my grandfather's family, I started wondering about their major life experiences and their impact. We all know people who've endured great traumas such as sudden close loss, combat PTSD, physical and emotional abuse, work injury, disease and other life-disrupting events. It is natural that those experiences inform our parenting now, and it's just as natural it would have done so for previous generations. For example, a mother losing a child to a hungry lion most certainly would be even more protective and cautious to be sure that no others were lost. While an objective evaluation of relative risks might produce one set of appropriate precautions, the wild unpredictability of the human mind and how it processes trauma might well lead to a different set of conditions. And some of these might seem a bit over the top, incomprehensible or outright crazy to someone who hadn't suffered that same loss in similar fashion.

I never knew my great-grandparents, but I have come to appreciate that much of my world was shaped by their gifts and their flaws, like personal relations. While Andrew the Fiddler was clearly musically talented, mechanically adept and bold enough to make an ocean voyage to a new world as a young man, it's worth considering the basics of his life and doing my best to understand them in context. He was the 6th child, in the middle of at least eleven children. He was the first of them born in Scotland, but lived with the stigma of being "Irish." Of his five older siblings, three were dead before he was born. By the time he came of age in 1882, a younger sister had also died in childhood.

I can't know how his parents might have coped with these losses, nor what resultant behaviors they impressed upon their children. When we found their grave in Scotland, they were remarkably silent on this topic and all others. My great-grandfather must have gone to great lengths to get himself into university rather than follow his father's footsteps into the mines. What part music played in any of his life choices, I probably will never know.

What I do know is that Andrew the Fiddler embraced that Scots heritage, as did his wife Margaret. She was born in Belfast in 1870, and emigrated with her family to Glasgow in 1873, and on to America in 1888. They both held fast to that identity, even though their family origins were in Northern Ireland. And once they were in

America, they seem to have integrated into the Scots community, leaving any trace of "Irishness" behind. Of course, as immigrants they still faced an uphill climb in the environs of turn-of-the-century Boston. They raised their children in the tiny houses they could afford, and do not appear to have been overly affectionate toward them.

At some point though, music was important enough to invest in passing it on to their children. While what I've pieced together is anecdotal, it appears that most of the children learned something of the piano and/or the fiddle. When they moved to Bridgeport in 1917, they bought a new house encompassing 1,200 square feet - with at least six children ranging in age from 6 to 17. My house here in Lincoln is cozy and comfortable for three of us at 1,400 square feet. It is no wonder that my grandfather and his siblings would have had far more than enough of each other by the time they came of age. One can only imagine that even if they were treated sternly at home, they spent a considerable amount of time unsupervised and out of the house.

Sadly, whatever childhood damage they carried forward into their lives had long-lasting repercussions with their own offspring. While Sarah maintained familial connection with her children and her grandchildren, most of the others were not able to fill much of any role as parent for any length of time. Of the other four siblings who had children that I either know or have been able to trace, none of them were able to maintain strong and loving connections to all of their children. The default parenting mode in our McKnight family seems to be one of estrangement and outright abandonment.

This is the world that Dad and Gramps struggled to navigate, as did I with my father when came my turn. With the benefit of 20/20 hindsight, I can see that my at-times challenging childhood happened because each of them chose to remain engaged with their sons. The odds were stacked pretty high against us having a "normal" family relationship, and overcoming that required a lot of sheer determination on the part of both Gram and my mother. For all of his flaws, Gramps didn't abandon ship or completely cut off either of his sons from his life. Staying engaged certainly left its scars too. But somehow between his choices, and my father's, that chain was broken.

Poor parenting does not negate the full power of your ancestors potential. And I don't claim that passed-down parenting from previous generations is among the most important factors in our childhood. Age and the life experience that accrues with it are a huge part of why our lives are different in our 20s than in our 40s.

When we are closer to our own childhood, we are perhaps more passionate about doing parenting differently than what we endured. I am grateful that I have been an older parent, even as I struggle to navigate my daughter's middle school world of social media and instant connectivity. At this point in my life, I have a different view than I did in my 20s of what to assign to the categories of big and small stuff.

But it is also inescapable that the full life experiences of each generation of parents - including their childhood - are intertwined into their parenting style and skills. When a substantial portion of a person's behavior and life choices are shaped by a significant trauma, it seems natural that might have impact the next generation and beyond. I'm guessing that I have experienced some of those repercussions for myself, even though I will never learn what might have precipitated them long ago.

> ***YOUR TREASURE: Why did they move? And just as importantly, how did they move? It's all too easy to restrict our view of our ancestors and their actions through the lens of our times. Life was certainly a lot different in the days before electricity or engines. If your ancestors suddenly picked up and moved to another area, how did they get there? Who did they know there? And what resource or opportunity awaited them that they were willing to make that effort?***

CHAPTER 18
Our Kitchen and a Family Meeting

Aunt Margaret the Opera Singer's grandchildren meet for the first time in our kitchen, and a wide circle at long last joins together.

Our house in Virginia was built in 1900, the same year my grandfather Andrew McKnight was born. Two stories plus an attic and a stone cellar, a small footprint but graced with a lovely covered front porch, and on a slope with a long narrow yard with plenty of room for gardens. It is humble, yet cozy.

Like in my grandparents' house, our kitchen is set off from the rest of the house - probably a prudent design feature in olden times when kitchen fires were the most common household hazard. Gram's kitchen was a pulsing center of the home too, but there were cabinets all around and anyone preparing food was isolated from whatever chaos was going on elsewhere in the house. In a home where lots of rambunctious children could interfere with preparing for a holiday meal, the isolation may have been a design feature as well.

Our kitchen was similar when we bought the house, and we opted to cut a hole through the wall into the dining room, and put a bar counter across into both rooms. And as such, the combined room became the "open" heart of our home, the gathering place for food, conversation and entertaining. In addition to the usual trappings of a simple kitchen and dining room, our framed family trees adorn the wall in honor and appreciation of our heritage - the known and the unknown.

So when we have guests, the food preparation area is the gravitational center of the house. During our annual Christmas party, it is alive with laughter, music and the mingling scents of many foods and mulled cider. Despite its humble appearance, it is where magic happens with our friends in our little cottage.

And on much rarer occasions, we host family. We live 150 difficult miles from my wife Michelle's Philadelphia family, over 400 miles from my family in Connecticut, and of course a long day's air travel from her family in Panama. It is a rare treat and treasure to have our kin in our home.

In the spring of 2018, spurred both by my fascination with Aunt Margaret the Opera Singer's family and the joy of connecting with her west coast descendants, I methodically turned my research attention to the daughter of her first child from her earliest marriage. Among the pictures I had seen for the first time in the past year was my great-grandparents holding the baby, standing next to their daughters Margaret and Sarah. Born in 1918, Virginia was their first grandchild. Since there was no father in the photo, I'd assumed that perhaps he'd taken the picture.

After six months or so, Margaret apparently filed for divorce and headed west - alone. Raised by her father, Virginia had married and reconnected with at least my branch of the McKnights by the 1940s. They lived in central Connecticut and Massachusetts, and I knew they'd had a daughter named Jane in the early 1940s. When we first connected with Aunt Margaret's eldest grandson Carson back in 2016, he shared pictures of his mother Martha visiting her half-sister Virginia during the late 1970s so I knew there had been some contact. I'd seen a couple black and white photos of Virginia with young Jane. But other than knowing she once existed, and that her mother knew of us, Jane was a completely blank slate.

Given my rich experiences of connecting with Carson and some of Aunt Margaret's other grandchildren, I got curious about Jane again. Within a few days, I was fairly certain I had found her, living barely 90 minutes away in the northwestern suburbs of Baltimore. But all I had was a phone number and an address, and a different surname than what I had expected. It was going to require taking a risk, making a call and hoping the phone didn't slam down in my face.

It didn't. A woman answered the phone, and I quickly explained who I was. She was indeed my second cousin Jane, who had lived her life with little knowledge and less contact with her mother's mother's family. We talked for a bit and she invited me to visit when I would be next passing through, in just a few weeks.

I was nervous about it, I admit. After all, I hadn't really met any of Aunt Margaret's family in person, and even with an invitation, here I was out of the blue about to literally show up on her doorstep. Having had a lifetime of experience with the finely-sharpened McKnight predilection for cynicism and suspicion, I was ready to somehow be disappointed. But I could tell she looked forward to meeting, as did I. She was intrigued by what I'd shared about our family's pre-Scotland roots in County Down, and my family's epic trip to Ireland and Scotland.

My worries were completely unfounded. On a humid early September day, she and her husband Ed made me most welcome. We chatted for a while over lunch, sharing some of the family stories we had heard. She knew her grandmother had other children and suspected she had cousins she didn't know about. And suddenly here at her kitchen table, it became quite real that there were people living who were connected to both of us.

It was after lunch when Jane brought out the picture that dropped my jaw. It was from my great-grandfather's 85th and final birthday in 1951. There he stood, while an unknown woman played the piano, with his fiddle up on his shoulder. It was the first and only photo I've seen of him playing music. And there, right in front of him, lost in the moment with a half-smile on her face, was young Jane. To this day she doesn't remember the event, but there she was.

I pulled out the gift guitar that had been made in our ancestors' hometown. It was then I realized that this beautiful instrument has become a symbol of our family in our 21st century, much like the coat of arms or a clan tartan that had meaning to our ancestors that is now lost to us. There in her kitchen I played her the first song I had written on it, "A Dram to the Holidays," a remembrance and appreciation for the hard lives of the ancestors who'd given life to us both.

We parted ways that day with warm embraces and a promise to not let too much time pass before we'd get together again. Little did I know how soon that opportunity would present itself. An opportunity of a lifetime, as they say.

It had been more than a year since we had connected with Carson. He had indicated for several months that he and his wife Valerie were hoping to be healthy enough to make a trip east someday. Not long after my visit with Jane, to my delighted surprise, he got in touch again to plan a visit for late October. I didn't realize until just a couple of weeks before they came that while they hoped to do some sightseeing around the city of Washington and the Virginia countryside, the main purpose of their visit was to meet their long-lost family, while they were still able to do so. So I was astonished and thrilled to find they'd be mostly available to us that whole week.

And now of course, they wanted to meet Jane, Carson's half first cousin. We all were most excited about a family union between the cousins who'd never met. We arranged a tasting at one of our local wineries followed by dinner at our house. As with most events keenly anticipated for considerable time beforehand, we all knew it would be over in the blink of an eye.

When the day finally arrived, we were treated to a Scotland-type day in the Blue Ridge foothills - low clouds and dreary mist mixed with outright pouring rain. No matter - the cousins all arrived and met at the appointed time. A couple bottles of wine appeared and the emptied vessels disappeared, fueling lively conversations and remembrances of different moments with shared ancestors. In a way it was the same feeling of astonishment and immense satisfaction that I'd had at dinner by the water in Northern Ireland, when our Irish and American families became one around a table of celebration.

Back at our warm and cozy kitchen, the seven of us all packed around our small dining room table and ate, while our combined pool of old photographs scrolled before us on my laptop. Those images had been held by our different families, handed down and somehow kept for all these years. The slideshow included a picture of three children - Margaret and Sarah as toddlers, and my grandfather in a baby buggy. To my knowledge, it is the only existing photo of my grandfather with his sister Margaret.

I've often imagined that Andrew the Fiddler must have played many of the same tunes frequently heard at Celtic sessions here in the States. The proximity of the northeast coast of Ireland to his Scottish childhood, the overlapping repertoire of tunes and the similarity of styles made it likely he would have known and played many songs we nowadays associate with Irish music.

After dinner my daughter Madeleine picked up her fiddle and played "The Star of the County Down" for her cousins all newly united, accompanied by me on the gift guitar. And just as in the picture from long ago, there sat Jane again, her face gently transfixed with joy as our fiddling ancestor's great-great granddaughter played. I felt like a man who has eaten the finest meal and would not be hungry for days.

I could never have imagined this blessing in my youth - that my personal interest in family history would bring these cousins to meet in their golden years, and that I could be present to experience those stories and memories. I am certain it would have been impossible for our great-grandparents, with those three little kids in the picture, to imagine us even existing 120 years in the future. And that we would choose to gather across 3,000 miles to dine together at a table because of them - or that those great-grandchildren would be the ones to stitch a ripped and torn family fabric back together across the miles and years, even as we continue discovering new colors of common threads.

I know that my great-grandparents carried their own flaws and damage with them, and sadly passed a lot of that on to their children. It isn't my point to judge them, only to learn what I can that helps me reframe my own story. I've always believed that each of us is worthy of some sort of redemption in the end. And if that's the case, I'd surely like to believe that they would be glad to be remembered and present at our table.

YOUR TREASURE: Modern genealogy is by nature a collaborative endeavor. When you work publicly on ancestry websites and collaborate with other researchers, over time and with scrutiny you may get a clearer view of your ancestory. You will likely have connections with people researching different branches of your family. Keep the basic genealogy in an easy-to-share format. Take care not to overwhelm new contacts with too much information. Help train them to do work in partnership at a pace they can handle, and you may well have acquired the most valuable treasure of all - someone to share in the journey and discoveries!

CHAPTER 19
Hourglass Women and Miracle Threads

Sometimes all of our everything comes through a single funnel.

My grandmother Madeleine generously shared her ancestral inheritance with her grandchildren anytime she could coax any of us to listen. Now I more innately understand her desire to pass that knowledge on, living with a tween in my house who has a passing interest in many things, and a deep interest in relatively few. She listens for a moment, filters what she hears from the entirety of what she's told, comments - or not - and is on to the next thing. I suspect I did the same thing to Gram, though I would try a little harder to wrap my head around my mother's stories about her parents and other people I never knew.

But my grandmother had better than good reasons to pass it along, and now I understand just how fragile our connection is to all her ancestry. Gram's younger sister, my great aunt Marjorie, was even sweeter than Gram and she doted on us kids since she had none of her own. I later learned that she had a hysterectomy at age 22, and thus was childless perhaps not by choice. Gram's descendants then would be the only ones to inherit their heritage.

Their mother Betsy Culver was also one of two children. Betsy's brother Harrison was cut down by typhoid fever at age 23, and so Betsy was also the only one of her family to pass on her lineage. Viewed through the genealogist's lens, my dad had no maternal 1st cousins due to his childless aunt, and no 2nd cousins from his maternal grandmother either.

I think of Gram and her mother as "hourglass women," not for the figures they might have cut in their youth, but as miracles of maternity. They were the sole bottleneck through which their ancestral genetic inheritance passed. All my grandmother's family lineage came only through her. Her two sons, six grandchildren, twelve great-grandchildren and so forth, carry that inheritance forward. But had it not come through Gram, that would have been the end.

My direct ancestors include several women who died young. One of those tenuous threads in my mother's family is my mother's maternal great-grandmother Alice Green. She married the son of Irish immigrants at age 18 and had five children. She died at age 30 while giving birth to triplets, none of whom survived.

Mom's grandmother Florence was only 10 years old when her mother died, so she didn't know much about her family and thus, neither did the rest of us. Florence married William Murray, who was also the child of Irish immigrants, and thus what we knew of our family history was the Irish three-quarters.

And that was that, until I started researching Mom's family. While Alice Green wasn't around to tell us her story, there was plenty of evidence to sift through. It turns out that my mother's Irish family also had deep colonial roots here in America. Alice's own mother had died at age 28. Through this mother and daughter, neither of whom lived into their thirties, my mother's family connected back to a lot of well-documented colonial ancestry.

Another of mom's 3G grandmothers died at age 21 after having two children. While it is impossible to not always think of her, and Alice Green and her mother being tragically cut down in their youth, I remind myself now and again how lucky I am that these "miracle threads" had children, and that those children survived to have children of their own.

Most behavior in the natural world favors passing on the strongest genes in a species. Evolutionary biology is full of case studies of various traits giving some individuals and/or species a leg or a wing up on those who didn't inherit those traits. Along with intense survival spirit, the desire to pass on one's genes is a primary driving force in the behaviors of most every species.

We humans are not immune to this urge. But we are uniquely able to create civilized societies with mores, norms and laws that govern and shape our behavior as individuals and communities, instead of by survival of the fittest.

I am raising my own hourglass, our only child. I've often thought about how each generation brings us farther from our past, but doubles our heritage. My daughter who wears my grandmother's name, inherited all my history and intertwines it with my wife's Spanish and Polish ancestry. Twice the tales that were lived, long ago and far away. She is the only one like her in the world.

Of course, every fiber of my being is dedicated to her surviving and thriving as she lives out her own best life. While I hope that her life will someday include children, I realize that there are myriad threats to that happening, simply in the statistically small hazards of everyday life. Along with all the other things in the wider world that I have no control over, I have to accept that I've already handed down my genetic treasure, and what happens to it going forward is largely out of my hands.

There are times when I am comfortable with that outcome, when the reasoned and logical aspects of the human mind are in charge. Then there are the other times - often in the sleepless hours of a long night - where I most strongly feel those primal urges to somehow have my DNA survive and continue to be passed on. I imagine that just about any parent would feel strongly to protect their child and find a way for them to have a long life full of their own children and grandchildren.

With luck some hundred years from now, her many descendants will see her life as a great cosmic thunderbolt of heritage, even as their own will be far wider and more complex than mine. Perhaps her descendants - who also come from my hourglass women and miracle threads - might even read these words and learn that their stories stretch far into a lucky and improbable past.

YOUR TREASURE: Time for a reality check about the big picture. We can't save everything, and the work will never be done. But when we do get a clearer picture of moments of our ancestor's life between the dry facts of their birth and death dates, take a moment to reflect on that life. What were you like at that age? What might it be like to experience the situation that you've learned about?

CHAPTER 20
Identity and Tribes

The mirror only shows us a piece of ourselves and our stories.

How did you describe yourself growing up? I don't mean the gawky kid with embarrassing features like big feet and uncooperative hair or skin, but the story of where some branch of your family originated? I've become fascinated by our identity stories and tribal affiliations, particularly as I've seen hundreds of people find their origin stories were in fact different from what they were told growing up. "You're a Scot," I'd hear from my grandfather, while Gram was proud of her original Connecticut Yankee heritage.

Part of our identity is often wrapped up in our traditions. Whether they come from our ethnic or religious identity, or simply were a thing that we did with our grandparents when we were kids, those are usually our most tightly held identity totems. Sometimes the tradition continues without any remaining connection to or knowledge of why it started.

Once Gramps passed away and Gram went into the nursing home, our family tradition of gathering at their house for Thanksgiving and Christmas ended, and it has become much more infrequent for my cousins and I to gather. Other traditions within our families with our own children have supplanted those and have become "what we do."

When I left for college, I was estranged from my parents for several months. It ended when Mom called to invite me home for a beef rib roast dinner on Christmas Eve. Perhaps she understood how easily the estrangements among McKnights could harden into permanence, or maybe it was the hole in our lives left by her parents' early passing. Whatever her motivation, she was determined that this would not be our story. It was no small expense or undertaking, and I recognized and accepted it for the olive branch that it certainly was. For these many years that my sister and I have made the long and arduous holiday journey to get back to our parents' house for the holiday, the rib roast on that night before Christmas is a tradition we've worked hard to keep, much like the relationships. To me, it is a cherished part of our family identity.

I have been impressed as well with how my daughter and her cousins hold fast to traditions that we started when they were little, like opening up their matching footie pajamas on Christmas Eve. In order to get a bunch of rowdy, noisy kids out of my parents' tiny house on Christmas afternoon, we started carving little scraps of wood into "boats," and then hiking a mile into the state forest to a rocky stream for our annual "Christmas Creek Regatta." For all of its silliness, it has become a tradition relished by the kids and their parents too.

Learning our family history bestows upon us a tremendous wealth of identity, good and bad. It is common to find some heinous people have contributed to our existence. We love to wear elements of our selected identity with pride, and keep quiet about the ax murderers and petty criminals, the ancestors who owned slaves, or those who committed wartime atrocities.

But we grow up with a tendency to treat those identities like they are etched in granite - a static representation of who we are. Often it is the viewpoint of one or more of our living relatives who share THEIR identity story with us, perhaps frequently. Given how quickly a family tree fans out as it doubles with each generation, it would seem that our identities are as much a product of chosen cultural affinities as they are of our ancestors' actual ethnic contributions to our genes.

Genealogy offers us a different lesson. Most of us come from a complex mosaic of tribes over the centuries. People changed religious affiliations and political allegiances. We weave together the stories of our parents' ancestors, and their parents' ancestors. Then when we see on a map and a timeline how and where they came together, often by chance or sheer luck, a richer and more colorful story usually emerges of our inherited identities.

I need to look no further than the tween living in my house. I have written extensively about all the stories in my family, yet these are only half of her family stories. For she is also heir to my wife's Spanish and Polish ancestry, carrier of unknown and untold stories from Fermoselle and Galicia, of Moors and Sephardi and Roma and who knows what all else, because we know so much less of her family history.

Those stories, those people's lives and struggles, they are all bound up in that same DNA, together with all of mine, the known and the unknown. While we often lament that we don't know more about our family history, the truth is we can learn a lot by studying the stories of where our ancestors lived during the time period they lived there.

We also must consider the big regional events that overlap their time and location. While most of us can likely recall a bit about the watershed events of the last couple of centuries learned in a high school world history course, many big events beyond wars and the Industrial Revolution potentially shaped our ancestors lives, and thus our own. Waves of 19th century migrations from Ireland, and late 19th century from Eastern Europe, and the Spanish flu pandemic that came home with the returning soldiers of World War I are but a few of the many big factors that may have influenced their decisions and outcomes.

The truth is our heritage is a gift because it allows us to personally connect to things we've learned along the way, even the stuff we don't want to embrace in our chosen identity. We don't like to think about the details of how Genghis Khan spread his genes so widely throughout his empire. But it is fascinating to know a name and a place from centuries ago and realize there may well be some direct connection.

When I learned the story about my 10-times-great-grandfather Nathaniel Turner and his tragic demise on a shoddy ship off the coast of colonial New England, I remember rolling the discovery around on my tongue with a sip of whisky, for a dark moment processing the whole notion of how awful it would be to drown in a shipwreck. And of course, it was my luck of the draw that he'd had children before it happened.

Hold that thought for a moment. Each and every one of us originates in the same way genetically - two parents, four grandparents, eight great-grandparents, etc. The farther back in time we go, in theory each generation doubles. So in theory each of us has 4,096 10G grandparents. (In practice we naturally come across more and more duplicate ancestors due to the small communities from which they came, a phenomena known as pedigree collapse.) Knock any one of them out of the tree, and our story goes blank.

That's a LOT of luck, isn't it? With all the diseases and dangers experienced by our forebearers, one could wonder how we got here at all. To pick out pieces of that vast tapestry to hold up and say we are more one thing than another seems a little less definitive in that view. Possible? Sure - there are plenty of cultures in the world where people have stayed in the same region with relatively little mixing for many centuries. They still had myriad ways to be taken out of the gene pool, yet in that one vital animal way, they were successful.

And thus, here you are. Glorious you, winner of that genetic lottery, champion over a bazillion-to-one odds to be here. Your very existence is proof that you are a miracle. But it also means that your identity story is rich and complex, with many pages and chapters. While your identity as Irish, West African, Thai or any one of another commonly discussed ethnicities may be true over a short period of history, don't sell yourself short.

While we are entitled to the leisurely pursuit of that identity selection, we are also heir to all of the stories by birth. We had no say in the matter of our conception or birth. Thus we have no claim to the triumphs of our ancestors any more than responsibility for their sins and failings. Those were their lives, and they are part of our stories - and only our stories. Although each of us along with our full siblings are the only ones who are heirs to our complete set of ancestories, none of those individually belong only to us. We may possess the pictures of that great-grandmother, but her stories are not ours to hoard. They belong to her whole family, of which we are a part.

Happily, all of that allows us to let go of any archaic notions like bearing seven generations of shame. After all, if you were the last of those seven generations, that ancestor's story might be hard to even find among the 64 chapters that generation contributed to your chest.

> ***YOUR TREASURE: Spelling hasn't always been as important as it is in our modern world, and thus surnames in your family tree could have several variants. It gets more complex when it has also been translated from a different tongue! This goes for place names as well as people - one region of my wife's ancestry has been part of Prussia, the Austro-Hungarian Empire and modern day Poland. And while people have been lying about their age for a long time, it is also possible that they really didn't know when they were born precisely. Perhaps the graveyard engraver needed something to etch in the stone, more than he needed to be correct for***

researchers several centuries in the future. As you go back in time, take care to consider clues you might be missing by assuming that your information is infallibly correct.

CHAPTER 21
The Graduation

Understanding the bigger family picture, and particularly the role of elders, changed how I make decisions about many things.

Revered elders are among the biggest stars in the screenplay of my life. How I grew to see these people in their senior years with that reverence, as opposed to a youthful inclination towards dismissiveness of "old people," is primarily a function of parenting. For my entire life, I have been the one looking up toward my favored senior citizens, appreciating their company and their candor - in particular the dearest and closest network of my parents' friends. It is simply how I was raised. I count among my own dearest friends several people fifteen to twenty years my senior. And since I don't have a corresponding cadre of much younger friends, I have truly never given much thought that someday with luck through a life well-lived, I might play that role in the lives of others.

My cousin Mike's oldest daughter graduated from Stevenson University in Maryland this past spring. While a few years ago I'd have surely thought about going, and may well have been tempted to do so, I probably wouldn't have made it to her graduation. The two-hour drive each way including Baltimore Beltway traffic and a litany of other plausible excuses would have allowed me to stay home and watch the live stream.

These past few years of family research have helped me better understand the primary people in it, as well as to better appreciate my own role in my time and my family. I don't have an abnormally-sized pool of regrets for past actions, but as I go forward I am hampered by having outlived my childhood imagination. I remember as a child thinking ahead to the year 2000 in the far off future when I'd be all grown up. I don't remember ever thinking past that point, which I've now outlasted by two decades. To the little kid in me, these are uncharted waters that I have to learn to navigate as I go. And that means looking at what's growing up behind me outside of my own house.

Despite the large families in much of my historical family tree, my sister and I have only four paternal first cousins and five on my mom's side. I am closer with some than others, as is natural with age differences and geography. At this point in my life, I am lucky to have only lost one.

Mike was the younger of Mom's older brother's two children, and a couple of years younger than me. My uncle was career Air Force, so their family were mostly just exotic stories to me because they lived far away and moved every year or two. When Mike's older sister Terri turned 18, she took a job in Connecticut and lived with my aunt's family until she could get her own place and settle there for good. The moving around thing was well worn out for her by the time she came of age.

And thus, Terri was the one I first got to know, before my uncle and his English wife retired and moved to central Connecticut too. Eventually Mike and his wife joined the rest of the family, and their two girls were born there. While it was cool to have their family closer while I was a young adult, we still didn't overlap much outside of obligatory holiday gatherings. Mike was always quiet too, so in the midst of boisterous family shenanigans with my other cousins, we never really had a whole lot of conversation.

I can only guess at some of the hardships in his life. It surely was tough to form friendships when the family moved about. Some of my uncle's deployments took him away from the family too. And my own experience with fathers and sons would suggest that even under the best of circumstances, that could be difficult for both.

Whatever voices haunted my cousin, they surely caused him great pain, and alcohol proved an unrelenting curse. When Mike got divorced, his wife returned to her family's home on Maryland's Eastern Shore and raised their girls pretty much on her own.

Unfortunately, cancer took my aunt away, followed by my uncle 15 months later. A month after his dad passed, and so many decades too soon, Mike's heart suddenly and unexpectedly gave out. His daughters were 13 and 10 when he died. I hadn't seen them since they were little girls around the Christmas table at my aunt's house, so despite my travels occasionally taking me past their hometown, we didn't share have any tangible connection. For his sister Terri to lose both parents and her only sibling in such a short period was cataclysmic. And suddenly she and her nieces became the entirety of her parents' living legacy.

The older I get, the more I appreciate picking up the phone and talking to my parents. I really can't imagine a tougher thing than a child condemned to growing up without one or both parents. Whether a parent's life is cut short or simply runs out due to the vagaries of age, the finality of it all is forever hard to grasp.

So when Mike's oldest daughter sent me an invitation to her graduation, I thought about it for a few minutes.

Since I had the day available and the campus was only about 90 miles away, I replied that I'd love to be there and therefore I would. I correctly suspected that no one else from her father's family would be able to attend, and I knew that simply showing up would make a lot of people happy - most importantly Elisabeth.

I did it because she did it. I got up and hollered when they called her name, as she made the long walk across the stage to receive her degree. It was a delight to spend time together over dinner afterwards, and be a part of their family celebration - my family.

What I've learned these past six years compels me to be around for my cousin's daughters in some meaningful way, nearly as much as I want to watch my own daughter grow up and have a family of her own. It was an honor to cheer Elisabeth on in person for such a major life milestone. Her handwritten mortarboard said it all - "We're all stories in the end." Her younger sister is following in her footsteps even as she blazes her own trail to graduation in a couple more years. I have plans to be there for her too.

But I have no aspirations to become a revered elder. That is a two part equation, and only one part of it happens naturally. It's not planned. The revered part is earned, in large part by giving unconditional love, by holding up a lens for the younger generations to see themselves in context of a larger story, and perhaps too, by showing up. I am happy to do that work solely for its own purpose and importance rather than how it might reflect back on me. My role is as a witness, and not a sage.

I am sure that my parents did not intend to become revered elders either, but that is their reward - the sweet fruits ripening in the late stages of lives well lived and well loved.

> ***YOUR TREASURE: What big family milestones have you witnessed? Funerals, weddings, graduations – they are all part of a family legacy. While having the dry facts one might find printed in a newspaper or online resource, a first-hand account can make that event come to life for someone a century from now. Take a few moments to collect your feelings and memories when you are among the witnesses to some family member's big life-event, and save it in your Internet safe spot.***

CHAPTER 22
Where Ireland, Cuba and Spain Intertwine

A week swirling with overlapping stories and connections deepens my understanding of my daughter's inheritance of ancestories.

Living generations make sure that my ancestral research isn't allowed to completely supersede the here and now, particularly the kids who quickly get bored with all the dead people stuff. Since my sister and I live so far apart, carving out cousin time for our kids is a stiff challenge. For the past several summers, we have brought our families to the White Mountains of New Hampshire for a few days of camping with our parents. We bring our kids to have some of our own childhood experiences like hiking in the mountains, cold river swimming, ice cream and of course, the rustic facilities in the same national forest campground where we spent many nights as kids each summer.

In the summer of 2019 though, tradition didn't fit so neatly into our plans. We planned our normal get-together with each of us celebrating milestone anniversaries in August, topped off by our parents' 60th. But my now 80-year-old mother had a months-long bout with dizziness and shortness of breath. She decided that she wasn't up for camping three-and-a-half hours from home and trusted medical professionals.

But summer rituals must be upheld somehow, and Aly was willing to pack up her kids and drive the five hours south to see us. We instead opted to camp in the nearby Pachaug State Forest, the same forest I'd spent so much of my youth exploring. My family would be spared the additional northbound hours. But best of all, now we were a scant 40 minutes from the Atlantic Ocean beaches my Panamanian-born wife treasures and so rarely gets to enjoy.

Our change in plans also facilitated some ulterior motives. In my 25 years as a recording artist, I had performed with my dad on piano many times. But we had never recorded anything together. Since this project draws hugely from Dad's family history, it seemed not only the right thing to do, but essential.

So I had written a song a couple weeks earlier, expressly tailored to his style of playing and his musical tastes too.

We recorded the basic tracks in Virginia and shipped them electronically to a local audio engineer in eastern Connecticut. Dad had a week or so to arrange and practice his piano and organ parts. Even better, the engineer would come and record in Dad's basement - that same music room where so much of my musical life had been shaped.

And there was an added bonus. I had wanted to record my sister playing piano for Andrew the Fiddler's song "Margaret," published in 1906, to use as the introduction for the album. But I had yet to find a way to record her on a decent piano near her home in the hinterlands of Downeast Maine. Now Aly was going to be right in the neighborhood for a few days, and we already had a recording session scheduled.

Sure enough, it mostly went off without a hitch. I bagged my studio objectives with little difficulty, ensuring that from my great-grandfather's song to my own child adding her fiddle and voice, five generations of my family would be represented. The camping was fine until our stay was cut short by mosquitoes. It wasn't just the swarms of them waiting to carry us away at all hours of the day. Acting with an abundance of caution when mosquitoes began testing positive for Eastern Equine Encephalitis, the rangers closed the campground the day before we were to check out.

Mom's cousin Bob Murray and I developed a deep long-distance kinship between our shared family research and our mutual love of hockey. At the end of our busy week, we had planned a day together visiting the resting places of some of my mother's Irish immigrant ancestors, as well as those of the colonial family they joined in marriage.

Our John Ryan and his bride Anastacia "Annie" McMorrow Kavanaugh came from County Carlow in Ireland. While we have yet to make a direct connection, both of their families in that area are historic Irish royalty. But by the time they packed up and sailed west in the wake of the Great Hunger, it is likely that they didn't have much of value beyond the people they left behind. They emigrated to New York around 1855, and within a couple years had made their way to the area between the villages of Marlborough and Colchester in central Connecticut.

Our Irish refugees managed to do well for themselves, farming the fertile hills of the Connecticut River Valley. By the first decade of the 20th century, two of my mother's widowed 2G grandmothers had accumulated a considerable amount of land in Glastonbury and Marlborough, now two of Hartford's toniest suburbs. By the first decade of the 21st century, anything of our family's accumulated real estate wealth had been gobbled up long ago. However, a great deal of Mom's maternal family history is recorded in the town halls and cemeteries there and in other nearby towns.

Mom and I met Bob and his sister Maureen that morning on a high hill cemetery in East Hampton overlooking the lake. We were blessed with a beautiful day to pay visits to our ancestors scattered across a 40-mile belt out to the hills back near Plainfield.

We shared stories and some pictures in between cemetery visits and backroads drives across central and eastern Connecticut. Thanks to some noteworthy colonial American named Carrier, the farmhouse John and Annie called home was already of historic significance when they bought it, so it stood handsome and well cared-for when we stopped for some pictures. Down the road a few miles, we found the family buried near some mysterious Ryans in the Catholic cemetery in Colchester. John and Annie's son had married one of our miracle threads, Mom's great grandmother Alice Green. When I had learned that Alice's family was a keyhole back into a rich and previously unknown colonial heritage, I discovered many of her ancestors - and ours - were buried in the Old Willimantic Cemetery.

Mom and I finally parted ways with our cousins at another hilltop cemetery, tucked away in a remote glacier-carved valley. Here was the final resting place Alice Green's father Nelson, a Civil War veteran whose family story had been quite a revelation. Before leaving, we paused a moment to reflect on what his parents must have felt when six boys of their nine children volunteered for Abe Lincoln's army. Four of them passed the medical exam and were pressed into service. Two of them gave their last measure of devotion on the fields of central Virginia, as General Grant pushed the Confederate army back toward Richmond during the final months. The war etched their family stories into smooth marble headstones.

While we had been busy camping, recording, celebrating and paying visits to our ancestors in person, there was much happening in my wife's family too. Though I have been fortunate that much of my own heritage could be traced several generations back with a nominal amount of effort, my wife's Polish and Spanish origins remain much more mysterious. Her father's Polish family tree runs into a brick wall with all four great-grandparents. Michelle's Cuban mother Queta was born to Spanish emigrants - a father from a tiny village in Almeria province near the Mediterranean coast, and her mother from the medieval village of Fermoselle, near the border with Portugal's northeastern corner.

Our Connecticut week coincided with Michelle's brother and cousins going to Queta's birthplace in far eastern Cuba to celebrate the 90th birthday of Queta's eldest sister, Michelle's Tia (Aunt) Eloisa. They went to the party and visited family in Santiago de Cuba, on what remains of the coffee plantation where my mother-in-law and her family were born and raised. We had heard of many of these places when Michelle's mother was still alive, but most of what we knew was from a few old pictures and what was told to us. Now we were enjoying the much-anticipated daily posts and pictures of living family from a distant place that we've never seen.

Ironically, when I started researching Mom's Irish ancestry, the first living relative in Ireland I met online was Maria Armenteros, Michelle's mother's distant cousin and not mine! Maria was born and raised in Fermoselle, but married an Irish artist and lived not far from where my family was in County Carlow.

It just happened that the same week my family was chasing Irish ghosts and Michelle's family was gathered in Cuba, Maria was back in Fermoselle visiting family. While she was there, she looked up records, talked to people and took pictures to share with us. And over the course of the week, a line of my wife's maternal grandmother's family unrolled before us, stretching back into the late 1600s. For a family story largely shrouded in myth and mystery, we suddenly had a clear picture of a new story, just waiting for us to dive into and explore.

So on our first day home from vacation, in between unpacking, doing laundry, and harvesting the overflowing bounty of tomatoes, beans and peppers from the garden, I immersed myself in the details of each of those experiences in succession. From Spain and Cuba came emails and posts with pictures of ancestors and long-departed relatives, all kin to my mother-in-law, wife and daughter. This started a flurry of looking up people and places, informally building a timeline of events, and revisiting still-mysterious connections between Spain, Argentina and Cuba in the early 1900s. They funneled through my laptop, processing pictures and taking notes to share with the appropriate groups of cousins.

We were experiencing the world of her mother and grandparents nearly first-hand through hundreds of pictures and videos, mostly showing family we'd never met and certainly of places we'd never been. There was the coffee plantation that her *abuelos* bought in 1946, left behind and lost after the Revolution, and now was being worked under government control by one of their grandsons and his family. We lingered over countless images of beautiful and rugged eastern Cuba countryside, and the stuck-in-1959 feel that still pervades much of the island today. And of course, there was the delightful video of spry Tia Eloisa leading the conga line at her 90th birthday celebration.

At the intersection of all these stories is our child. When I finally turned in, hours after midnight it hit me hard that she was the only person in the world who inherited all those stories as her birthright. Our hourglass is the single heir to the entire legacy of my family and my wife's family. Each of our lives was imprinted indelibly with hundreds of elements of these stories. They are part of who each of us is, and we are the people who are shaping her - for better or worse.

It is one thing to know the bare bones of your origin stories such as the dry facts of when and where some of your ancestors lived, and maybe their vocation or religion. For any person, that would be a terribly short summary of "the dash" - that mark on the gravestone between the birth and death dates that represents the entirety of a person's existence. In our family, that dash might be more like an arrow trailing threads of many colors in its wake, tracing an arc of exile in a diaspora from Ireland or Cuba to a new beginning as a hyphenated-American. For my Cuban-to-the-grave mother-in-law as much as my Irish ancestors, theirs was a land of no return, even if their descendants were able to visit again one day.

It is that lens which immeasurably deepens our human experience, as we look at their pictures, learn their stories or hold objects that were part of their lives in our own hands. To visit their homeland, to witness their sunsets and feel the heat of summer, to taste their food, to hear their language; these visceral connections to who they were tell us much about HOW we are.

Our inheritance is rich indeed!

YOUR TREASURE: While we might be lucky enough to have one-of-a-kind family treasures, it's worth considering that for our long-deceased ancestors, their photos and letters really belong to all of their descendants. Some people get angry when they see pictures from their collection attached to that same ancestor in another person's tree. But our ancestor's story doesn't belong ONLY to us, it's really part of each of their descendants. As with most things online these days, if you really don't want it shared, it's maybe better not to post it in the first place. But take a moment to consider your ancestor from your own perspective – wouldn't you prefer to be remembered by all your descendants long after your journey here ends?

CHAPTER 23
A Final Farewell to Aretas Culver

Walking in his footsteps at Antietam led to an unexpected quest for a gravestone and a very special Veterans Day ceremony.

I am standing on this windswept hilltop at West Cemetery in my dad's hometown of Bristol, Connecticut, in the shadow of the tall monument that honors the city's Civil War dead. Tomorrow will be the 11th of November - our American Veterans Day, established to commemorate the Armistice that ended World War I, the "War to End All Wars." But present on this day in 2019 are family members, dignitaries, veterans, historians, curious onlookers and a reporter from the *Bristol Press*. Today we are here to correct an oversight, to take care of one remaining piece of unfinished business.

It has now been five years since I first connected the dots and met Aretas Culver face to face, as his eyes stared back at me from the picture on my laptop. To this day, I've not seen the original photograph, nor do I even know where to find it. I certainly could not imagine the journey I would be making with my three times great (3G) grandfather since that discovery.

I am one of Aretas Culver's many descendants. He had a family, buried a wife and started another family before enlisting in Abe Lincoln's Union Army in the summer of 1862. He was thrust into fire and failure at Antietam, and 19 months later captured and held prisoner at Andersonville. His captivity would eventually cost him his life.

It is human nature to be drawn to tragic events, particularly when they intersect directly with our ancestors' stories. Since I live so close to many of the epic battle sites of the Civil War, it was possible to learn more about Aretas' experiences - a lot more.

I live an hour from Antietam National Battlefield, and I've visited many times. The disconnect of the deadliest day in American military history happening in such a pastoral Maryland setting is always deeply jarring, as well as profoundly moving. But I hadn't been to the battlefield since I learned Aretas' story, and thus had no idea of the critical role his unit played.

The 16th Connecticut Volunteer Infantry had no real hope of success as they marched across those steep ravines, new untrained and untested recruits facing a sudden charge on their flank from the Confederates' most experienced army. Yet many of the survivors spent much of their post-war lives trying to clear their names of accusations of cowardice on the field, as well as blame for the three years of war that followed. Their failure in that cornfield, while completely understandable in the calm hindsight of the historian or military strategist, hung on these men long after the guns of war fell silent.

I had never explored the battlefield around the Final Attack, where the Confederacy had staved off its demise. As overwhelming as other parts of Antietam's features can be, one is often saturated by that part of the tour, and the Final Attack sites aren't close to a road. I hadn't known of the monument erected to the 16th Connecticut many decades later by the state of Connecticut.

On a cold January day under a brilliant blue sky, I returned to Antietam along with my friends Prescott and Ian, who provided a rich trove of humor and military knowledge. Neither of them even had ancestors in America at the time of the Civil War, so they were thrilled to join me on a mission of intensely personal significance. The park rangers were all too happy to share maps and troop documents for their regiment, and a short visit with them at the visitor's center gave much deeper context to what happened to them that day. The three of us did the battlefield driving tour, but our whole visit was weighted by the anticipation of what would come at the end - walking in my ancestor's footsteps from the Burnside Bridge to that fateful cornfield where so many had their first and last horrific taste of battle.

As we hiked across the narrow valley and up the hill to where the obelisk marked their forward progress, I tried to envision the rising dust of several thousand men quickly approaching. It is only when the following columns come into view that the colors of a deadly enemy would be visible. But by then it is too late. We stood at the monument and studied its depiction of Col. Newton Manross, who fell vainly trying to rally his frightened and inexperienced neighbors.

It is unspeakably powerful to walk in the footsteps of your family, and to experience ghostly visceral wisps of what they might have seen, heard and smelled. All I could really do was to take in the incredible barren beauty of late daylight on this rolling landscape. There were no sounds of roaring cannons and dying men, nothing but ghosts, crows and corn stubble.

In November of 2017, Dad and I made a trip together to his boyhood village of Forestville, long ago swallowed up into the city of Bristol. It was a few months after our epic family trip to Ireland and Scotland, and our mission on this day was to deliver the soil we had brought from his great-grandparents' graves. Dad has never had any interest in the Civil War, but by this point even he had gotten caught up in Aretas Culver's story.

Once we scattered the soil over his grandparents' grave, we wandered about the Forestville Cemetery. There is a large group of Gram's family graves, a few yards away from the final resting place of Colonel Manross. Just up Central Street is Dad's boyhood home, across the street from the Manross Library. To say that Forestville is full of memories for him would be an understatement.

I met cousin Mike, a fellow Culver descendant and lifelong Bristol resident, through Ancestry. Over the months he and I shared a lot of research about Aretas' ancestors and descendants, and been fascinated by many elements of his story. We left Forestville and headed over to the West Cemetery to meet Mike in person, visit Bristol's memorial to its Civil War dead, and perhaps find Aretas' final resting place too.

It didn't take long to find Aretas' name, inscribed in stone alongside his many fallen comrades who'd given their last full measure someplace in that journey between Antietam and Andersonville. Many of them lay nearby under beautiful marble headstones with their rank and service neatly noted. Sadly, we could not find a stone for Aretas, though other Culvers were buried nearby. Dad and I parted ways with Mike and thought contentedly we had accomplished a lot in our long day's visit. On the hour-plus ride back to Plainfield, Dad shared a ton of stories about his childhood, another rare and unexpected bonus.

Mike and his father (and fellow Aretas descendant) Ed came to Antietam in the spring of 2018, and my daughter and I met them there. We again retraced the steps of the 16th, until Mike took an inadvertent step and broke his ankle. Although sadly our visit was cut short, he could now claim the experience of being carried off the field wounded at Antietam, albeit under far less harrowing circumstances than his townsmen 156 years earlier.

He was convinced that Aretas' final resting place was indeed at West Cemetery somewhere, and he stubbornly resolved to find it. Not long after his recovery from his unfortunate misstep on the battlefield, Mike's methodical searching and eagle eye located a simple broken stone next to Aretas' first wife and my 3G grandmother.

That broken stone bothered him considerably, and he wanted to do something about it. I was pretty sure that the Veteran's Administration would provide a proper headstone for any vet who had served honorably. Certainly they should do so for a Union Army veteran whose service cost him his life.

It was early 2019 before we'd revisit the subject. With a gentle nudge from Mike, I quickly found the VA application for replacement headstones for service members. Sure enough, we could pick the style and text on the stone, and after a few months they would provide it directly to the cemetery. It seemed fitting that he have a period tombstone, like the ones that honor his fellow soldiers from Bristol and line the national cemeteries at Antietam and the other battlefields of a most uncivil war.

A century-and-a-half after the surrender at Appomattox Court House, there remains a huge and impassioned interest in the devastating conflict that tore Aretas Culver's nation apart. I sensed perhaps my grandmother, were she alive and knowing all that we now knew, would have made sure the right folks in the right places knew about our little project. This included her beloved Bristol Historical Society. It seemed worth at least a bit of outreach around Bristol, and perhaps there would be some local interest in our odd journey.

It turns out there is a Veteran's Office right in City Hall, and I quickly connected with its head and heartbeat, Donna Dognin. Her duties include regularly coordinating events with West Cemetery, as well as the local veterans organizations. Before we knew it, we had a plan including an honor guard, a speaker from the Historical Society and the city historian. It all added up to a proper dedication ceremony, and it only made sense to do it around Veterans Day.

I have been fortunate recently to work with combat veterans burdened with PTSD, in a songwriting-as-therapy environment. Through this one-on-one work of helping them reframe their stories of trauma from an outsider's perspective, I have learned a great deal about the details they share reluctantly with those who have not been in the line of hostile fire. Thanks to their openness and my overactive imagination, many of their stories have become all too real to me. It is natural that the lens through which they have helped me see their experience would also bring more vivid insights into my ancestor's experience.

One thing can surely be said of war - the technology of killing is always a step ahead of the evolution of strategy and tactics. The horrors experienced by our vets in Iraq and Afghanistan are likely unique to their time, just as major advances in the range and lethality of Civil War weaponry was surely an unforeseen and devastating trauma for those who endured it then. As I reflected on the topography and circumstances of their positions at Antietam, being surrounded and shelled at Plymouth, and incarcerated in the front parlor of hell itself at Andersonville, I can't begin to imagine the impact it would have had on a man with five children at home, who'd lived a simple life in a small town. It would seem a complete and total sacrifice of everything cherished. And rather than a quick death on the battlefield, it became an ever-increasing weight of horror underneath his kepi every step of the way.

And so it all converges here today, at this cemetery in a town where I never lived, to honor an ancestor who I never knew and who died 100 years before I was born. Who passed his DNA to me through two generations of hourglass women who each lived long enough to have children. An ancestor who volunteered to save his country and witnessed horrors I could never fully comprehend. He died a full ten years younger than the age I am now.

We have now dedicated a proper memorial to Aretas Culver, one that acknowledges that last full measure he gave to preserve a nation that nearly tore itself apart over the right to enslave other human beings. He lies within sight of the memorial that acknowledges the sacrifices of all Bristol's fallen. I never did find his original stone. My cousin Mike did that. But as he points out to me now, the eagle perched at the top of the monument looks directly towards it.

We are seated on a rare level patch of ground, between the hilltop and Aretas' bright new gravestone. Three elderly local vets comprise the honor guard. They and the eagle are silhouetted against the grey sky as they fire three times into the air. One pulls the bugle and plays an elegant rendition of "Taps." I have no way of knowing how many times this scene was repeated between the aftermath of Antietam and the end of the uniformed butchery some three years later. I do know I am emotional hearing it, feeling both the gravity of the moment we are witnessing, and that these moments represent the end of a long journey.

As we walked back to the car after the ceremony, the interview and posing for the requisite pictures, we pass at least four other headstones of Aretas' fellow veterans from the 16th. One of them never made it out of Andersonville, and another did and lived until 1909. The randomness of fate with young men and war is as it has sadly always been so. In this moment I find it oddly comforting that Aretas Culver drew his last breath at home, surrounded by family.

Perhaps my grandmother, in whatever spirit life awaits us, has known this the whole time. She may have been pulling some great cosmic strings to set these events and people in motion. It is likely I'll never know, certainly not in whatever time remains on this earthly plane.

But on that hilltop cemetery in her hometown, I closely felt Gram's presence. She always brought flowers to the family graves and honored the family veterans, particularly those Revolutionary War ancestors. Were she in my place, she would certainly have taken the time and effort to procure a proper stone for her great-grandfather.

She made sure that I did my part to keep that legacy too.

> ***YOUR TREASURE: Whatever you do or don't do, you could make the case that you'll never know less than you do right now. You have the ability to tell your ancestories in whatever fashion you see fit. They can inspire historical fiction, or you can simply augment what's known about them in a free collaborative environment like Family Search or WikiTree so that future generations might know them as you've grown to see them.***

Epilogue

What a trip it's been, and who am I now that I am here?

On this frigid late November afternoon, I am again taking my daily walk down the winding gravel road. The leaves have carpeted the ground in their rusty hues, and the yellow-gray cornstubble stands knee-high in the fields. The birds of summer have long departed, leaving the cardinals and bluebirds to bring some visual relief from the dreary November landscape, along with a lonesome few Osage oranges clinging to naked tree branches like unripe lime-green brains.

The annual blanket of vegetative dormancy settles across the Crooked Run Valley, broken mostly by the unfettered northerly breeze and the occasional motion of critters getting on with the business of getting ready to endure the winter. As our magnificent globe tilts toward the winter solstice, the sun is already teetering low-angled in the sky at mid-afternoon, but it is still not as dark and as low as what my ancestors would see in the north of Ireland.

Much has transpired over these many days and seasons of walking my historic village and its old sunken pathways. My walk is no longer what it once was, for I am changed - not by the seasons or the years, but rather a different sense of understanding of the way the world works and the ways of human connection.

I often think of these past six years like a PhD research and dissertation. It is fitting that this story ends someplace entirely unexpected, for this is the way that much of it has unfolded for me. I've spent that time doing the work, but by far the most enjoyable element of it is the unexpected and unlikely experiences I've had along the way. Often they happen simultaneously, on parallel yet completely unrelated tracks, which gives an even more "You can't make this stuff up" undertone to the entirety of the experience.

Like many, I struggle with mood during those shortening days of low-angled light. It is only alleviated by the brightness of sun on snow, or a crystal gleaming winter moon. I try to imagine my ancestors in their time before electricity and in their worlds, all living in more northerly latitudes than my Virginia home. I wonder how they managed. Chasing down my ancestors and their stories has been one of many welcome outlets to bring light into my November darkness.

And thus I have come to regard these looming long nights of winter's onset as high genealogy season. I often think of Gram, and lament that we are not able to share in this journey together beyond wrapping her afghan around my shoulders. Perhaps it is that night seems to last 16 hours, or maybe it's the influence of an insufficiently wee dram, but I have often felt as though she has been with me in some way from the spirit world.

People speak of the "veil being thin" at this time of year. It does feel like the boundary between the earthly realm and what lies beyond could somehow be more easily crossed. It seems that many cultures have some ritual observation that includes ancestral remembrances during the weeks of sprinting into deepening darkness toward the winter solstice.

Most ancient cultures - at least those sufficiently far from the equator as to have seasons - seem to have ritualized the summer and winter solstice in some way, and other solar interval points like the equinox. In the ancient Celtic world, *Samhain* ("summer's end," in Irish Gaelic) divided the year between the lighter (summer) and the darker halves (winter), and marked the end of harvest season. Catholics around the world observe All Saint's and All Soul's Days on the first two days of November. My wife's many years of teaching Spanish meant we always looked forward to the colorful celebrations of Dia de los Muertos (Day of the Dead).

While I'm not a particularly religious person, I do consider myself deeply spiritual about our human connections to each other and to the mysterious unseen in the universe, as well as what is around us in plain view. I was not raised Catholic, so All Soul's Day is not a tradition with which I grew up. It is only in these past few years as I've become my family's "storytender" that it has taken on meaning to me.

A few years ago some dear friends invited us for a Sunday afternoon tasting at a local winery. It happened to be All Soul's Day, and I was just starting to make some of my discoveries about my family history. I suggested for fun that we should each come ready to share a story of an ancestor, raise a toast in their honor and say their name out loud.

Ours has now become a cherished tradition, borrowing bits from Dia de Los Muertos and Samhain, and mostly for the simple pleasure of sharing a little family lore with each other. Each year, our little celebration of this liminal time comes infused with new layers of emotional significance, as we get older and as more of our elderly relatives pass on. We bring pictures of our ancestors, light candles, tell a story and of course, enjoy the excuse to eat and drink together.

I've come to think quite differently these past six years about the dead, and about life and death itself. Some of that is no doubt attributed to being in my middle years, and closer to the end of my own journey than the beginning. It also comes with the natural progression of being a parent and wanting to be there for your progeny for as long as you can.

In the spring I was asked to help eulogize my friend Chris King. He was many things in the 20 years I knew him - devoted husband of his beloved Dene, past President of the Unitarian Universalist Church of Loudoun (UUCL), actor and arts supporter, local radio host and programmer for many years, and much more.

Chris invited me to play at the dedication of their little country chapel back in 2000. It was built in 1890 by the formerly enslaved people working on and around Oatlands Plantation and dedicated as an African Methodist Episcopal church. For nearly a century, those families and their descendants gathered on Sundays and soaked into those wooden walls their joys and sufferings from life during segregation, lynchings and Jim Crow laws.

When that congregation outgrew the chapel, Chris and Dene helped facilitate its purchase by UUCL. On the morning of the dedication ceremony, I distinctly remember the gloominess of the day and how partway through the service the sun seemed to explode through the window of the chapel. It felt sacred, and the timing was surely otherworldly.

Chris intimately knew and loved the blues, both the music and its history, and found commonality with the struggles and suffering in his own way. At Dene's funeral service six years ago, the look in his eyes told me that a big part of him went with her and was never coming back. Maybe this spring she opened some cosmic door for him, and he slipped on through.

As I thought about what to say at Chris' memorial service, I knew there had to be some touch of the blues, perhaps on the slide guitar, maybe evoking the iconic Delta bluesman Robert Johnson. But it wasn't until that morning "When My Time Comes" emerged from the ether over about fifteen minutes time, raw and unpolished. It seemed fitting to send him off not with a lamentation, but a celebration of life. As we sang Chris farewell with one of his favorites, "Will the Circle Be Unbroken," he and Dene sent the sunburst through the chapel window for us once again.

My time is now. It's time to live to the fullest, to be present with my aging elders and with the generation following in my footsteps. It's time to take note, leave stories and drink in all the richness in between. There is much of all of that yet to be done.

If we are lucky enough to be blessed with old age, it is likely accompanied by a blessing of bewilderment as we gaze out on a world we barely recognize. With each visit back to my parents hilltop home, the old relic stone walls of my childhood are disappearing as the land around them is developed. Those stones fetch a pretty penny on the landscaping market in wealthy suburban New York communities. We may lament the loss of people, things, traditions and places as much as we cherish the progress we've witnessed. That loss creates a deeper connection to some of the things we've held onto as well as the memories.

I have learned a great deal about who I am. While I am the singer/songwriter and musician from Lincoln, VA, born and raised in Plainfield, CT, there is so much more to the stories and possibilities I have inherited. I am content with what I have learned and accomplished, but I can also mostly satisfied with the much larger pile of what I don't know. We can let our imagination fill in the blanks, or we can simply let it rest. Our inheritance is here safe within us.

My ancestors were part of the waves of those fleeing religious persecution in England, who founded towns and stole life and land from the native inhabitants of coastal "New England." My people braved boats over angry seas for many weeks in search of something better. Some fled famine and hopelessness in the heart of Ireland, while others were run out of Scotland long ago for sheep rustling. They aimed weapons in anger at the English in the Old World and the New, and gave their lives to keep their piece of the New World intact in the hope of someday achieving a nation with all created equal, and justice for all. They worked in mines and mills, and on boats and docks. They were farmers, weavers, clockmakers, musicians, preachers and of course, mothers and homemakers.

I am the great-great-great grandson of Aretas Culver, and of John Ryan and Anastacia McMorrow Kavanaugh. I am the great-grandson of Andrew McKnight, and a grand-nephew to his daughter Margaret the opera singer. I am second cousin to Carson, and Melody's third cousin, a third cousin once removed to Paul and the fourth cousin of Eileen, and I am probably Cathy's fifth cousin. My people come from places like Castlehill and Newtownards in the north of Ireland, and from Glountane and Sherwood in the south, from Nortondale and Woodstock in New Brunswick, Canada, and from Bristol, Connecticut.

I am related to thousands of names I know, and literally millions more that I never will. And while we share pieces of this story with them, my sister and I, and our children, are the only people who share all of it.

And of course, I am me. One in a bazillion, one of a kind, endowed with so much of the strongest genes of my ancestors, entitled to none of their accomplishments or failings beyond those worth recounting in a good tale over a better ale. I am just like you. Each of us are heirs to a vast and magnificent repository of unknown stories, distant heritage and more mysteries than we can even imagine.

Most of all, I am the proud and lucky grandson of Madeleine Warner McKnight.

Acknowledgements

As with all epic projects, this one was made possible by the cooperation and complicity of countless people, including a great many to whom I am related. My parents Kathy and Warren who went along with these shenanigans; they took DNA tests and wound up going to Ireland and Scotland. How cool is that! They and my sister Aly have all encouraged, aided and abetted me in countless ways in this journey, for these are their stories too. I hope this does them justice and brings joy as well.

My wife Michelle has put up with more than most saints in tolerating me and my crazed, occasionally obsessive creative pursuits. Like my parents, she has never doubted my ability to bring this whole thing to life, despite neither of us truly having any notion of how to do it. More than anyone else, she is innately familiar with many of the stories and experiences chronicled in these pages, and yet she cheerfully served as a beta reader as well as a sounding board.

Quite simply, this wouldn't have happened without Chris Nicholson's nudging and encouragement. He spent untold days editing and challenging me to focus on clarity and relevance without losing the stories, including not letting a good Titanic story go to waste. I will be forever indebted to him and to Maria de los

Angeles, whose kindly spiritual wisdom and vast writing experience helped shape this into a much better book. Her deep influence might well be summarized in a bumper sticker slogan; “less academic, more emotional.”

Chris is also one of my mastermind group along with Rich Wilkinson, Mel Harkrader Pine and Brian Giblin. True and longtime friends all, willing to come together at my table and brainstorm about how I might turn my passion and my art into a viable and slightly more comfortable income than the typical gig to gig economy of most touring singer/songwriters. We're still working on it, but the dinner and libations around my kitchen table have produced a ton of laughs, and also a lot of great ideas that have helped me "move the needle".

Larry and Betty Nilson, like a second set of parents to me, but free to be my lifelong champions without having to worry about disciplining an unruly child :) Two people who have believed, encouraged, cheered, lent a hand and made time for me like none other. I am forever grateful for how generous they are with their love.

My family, many of whom I am closer to now than I ever have been, and that in itself is reward enough for this project. My aunts and uncles, present and departed, and my nine first cousins in this world are the family I've known all of my life, and my great Aunt Muriel Tarwood and my late great Aunt Phyllis Arnold who kept these stories for us deserve special mention and affections.

To my living cousins with whom I've actively shared this journey here in America - Bob Murray, Mike Quish, Chuck Hasbrouck, and my McKinfolk - Carson and Valerie, Lee Johnson, Karen Platts and all of our shared family. And my Irish cousins Paul Simpson and Eileen Murphy, Cathy Callan and her dearly departed mum Margaret Fenelon, and the Scots too - Kerrie Drysdale and Daryl Rodger. I'm especially grateful to Leah Cooper for all of her work separating fact from myth about our Titanic ancestor Charles Kirkland, and to Hazel Fawcett Fekete for her collection of Nortondale history.

Special thanks to each and every one of my cousins close and distant who took the DNA test and in someway enabled me to learn more about our past! In the work of better understanding and properly using DNA - Blaine Bettinger and Jason Lee are two of the many creative and enthusiastic researchers who share their talents and techniques with those of us seeking to make use of all our evidence! Special thanks to my dear friend Dr. Joe Angel in Austin for kindly reviewing my basic cell biology and genetics discussions so that I might speak truth in plain language for the curious.

For all the people with whom I've connected with in my research, whose own passion and skills have in some way enabled my work and made my way much clearer. Those include my dear friend Lea Coryell who guided me through a wonderful research visit at the National Archives, as well as Howard Mathieson's generous heart and brilliant data and mapping skills, and Greg Campbell, the local historian in Woodstock, New Brunswick. I am also quite grateful to many genealogist/ teachers in the "public square" online and as speakers, in particular Amy Johnson Crow, Crista Cowan at Ancestry, and Maurice Gleeson.

The Aretas Culver part of my life probably would not have happened had I not had the good fortune to find the published work of Dr. Lesley Gordon. What we have learned through *A Broken Regiment* is a gift indeed, as are the contributions of Donna Dognin at the Veterans Strong Community Center in Bristol CT and the staff at Antietam National Battlefield.

For the work bringing the music to life, I owe many large debts indeed. To Dustin Delage for recording it and the many hours sculpting it into fine art, to Lisa and Rachel Taylor, Michael Rohrer, Jeff Arey, Jon Carroll, Tony Denikos, John Rickard, and my longtime bandmates Stephanie and Les Thompson for adding their talents to the music. Bill Wolf is my trusted "finisher" for making audio sound great, and Aimee Weakley at Oasis Duplication deserves kudos for getting my projects on disc for those wonderful folks who still buy them. Special thanks to Michael DeLalla for being my sounding board from the conception of this whole crazy idea to this finished pairing of words and music you now hold.

The graphic design talent and skill of Stilson Greene graced every aspect of this project, for which I am forever grateful and blessed. Christi Porter's keen lens and innate sense of "me the artist" similarly helped bring the visuals into sharp focus. Many thanks also to Dave Levinson at Wicked Design for his kindly guidance at bringing vital marketing elements together.

A lot of folks gave generously of their heart to make this crazy notion evolve into these songs and stories.

Priceless Treasures – the Mitchell and Langley families, Eileen & Mark Murphy, the Langley Family, Scott & Becky Harris and the Catoctin Creek Distilling Company, Elizabeth Krouscl Simmons & Steve Simmons, Betty & Larry Nilson, Bill Sergeant, Tony & Susanna Pangilinan, and Anonymous Angels.

Golden Treasures - Don Bunker & Becky McEnroe, Lori Rock & Pat Crotzer, Leslie Murphy, Nancy Clark, Sarah Harper, Amy, Stephen & David Brand, Cindy & Lani Pearson, Jesse Gallagher & Seery Strings of Bristol CT, Joan Clark, Tony Denikos, Laura Solomon.

Cherished Treasures - Jane Bagby, Bob & Doris Murray, Marjorie McKnight, Susan & Danny Anklam, and Bev Turner, Mary Chapman Cole, Laurie Grekula, Mel & Carol Harkrader-Pine, Betsy Dake, Brian & Susan Giblin, Peggy Fallon & Bill Schell, Paula & Scott Moore, Bob Sneed, Chris Nicholson, Susan Couture, Eva Belt, The Schimmoller Family, Annie Jenkins, The Mills Family, Leslie Solitario, Tamar Datan & Sandy Shihadeh, Jenna & Wyatt Korff, Frannie Taylor, Elizabeth Menist, Jim Dake,

Kevin & Jaynne Keegan, Reverend Jo VonRue, Debra & Paul DeKeyser, Tom Cardarella, Barbara McKee, Chris Haines, Jeff Edgin, Paul Laughlin, JoHanna (J) Whitfield-Brogan, Dan Tappan, Terrie McClure, and in loving memory of Jack Drucker.

My musical life has been greatly enabled and abetted by many worthy of deep gratitude, from giving me safe harbor far from home to helping bring my music to the world. Special thanks to our Mountville Folk Festival and UUCL communities, and dearly generous friends Eleni Kelakos, Sue Gaines and Sarah Huntington.

Finally, to my dear cousin Melody, who shares a big part of my grandmother's lineage, has steadfastly believed in this project and the power of these stories to help others, and backed it up with her Spirit Ninja Publishing company. With love and a toast to our "hourglass women!"

And to Hazel and Perley, the grandparents I never knew. I'm pretty sure you'd have thought I was a swell kid too.

APPENDICES

Appendix A
Family Connections

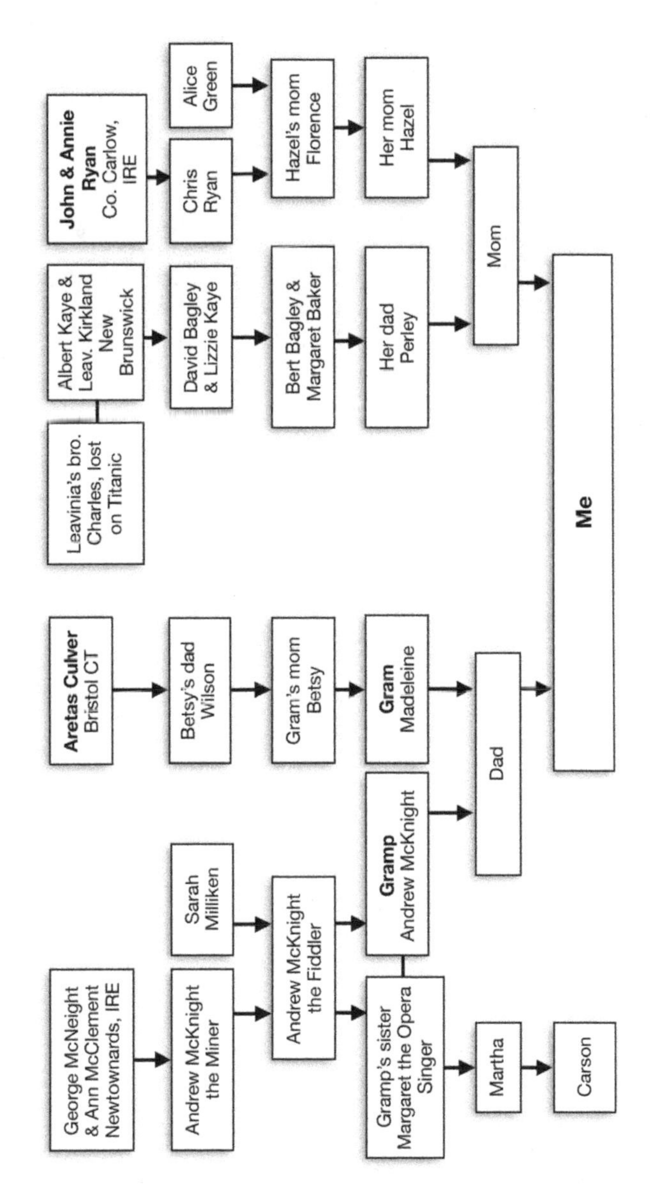

Appendix B
About the Recording

The book comes with a companion music album. If one is not included, you are entitled to download it for your personal use and enjoyment - please email info@andrewmcknight.net for download instructions.

Lyrics, recording credits and liner notes below.

1. Embarking

2019 ©Catalooch Music, BMI

A slightly Celtic sounding melody created on a Native American flute, and set on the western shore of the North Atlantic. With thanks to my wife Michelle and her iPhone for recording the waves lapping along the RI beach, and to my friend Eva for the precious gift of the flute.

AM - cedar flute

2. Margaret/Treasures in My Chest

1906 © Andrew McKnight, Success Music/2019 ©Catalooch Music, BMI

This project took a big leap into reality in the spring of 2016 when my parents introduced me to their copy of "Margaret," published in 1906 by my great-grandfather whose name I happen to share. It was a profound experience to find an ancestor who also "hears the voices" and was also driven to turn them into verse and melody, and that experience became the seed to "Treasures in My Chest". It wasn't until I connected face to face with some of my musical second cousins in autumn

2018 that I realized that visit was what I needed to finish the song.

Aly McKnight - piano
AM - vocal and guitar

Scent of cedar rising, as I open up the lid
childhood things and so much more, darkness kept safe hidden
and it holds my stories like a sailor's prized possessions, while he's out to sea

Wrinkled scraps of paper, and faded photographs
my name on a manuscript, and notes upon the staff
the lines unfolding on these pages I am holding, written long ago
are the gift running through my veins, and fingers on the strings as they flow

> From these treasures in my chest
> timeless rhythm beating, steady in my breast
> memories and hand me downs present from the past
> countless treasures in my chest

Pieces of this puzzle, of how I came to be
family names and question marks, somehow part of me
now I'm climbing up this tree exploring, branches that I see for the first time
samaritans and saints, rogues and renegades, all come alive

Message tone awakens me, from my imagining
connections in the here and now, that music brought to me
for they've sat here at my table, raising toasts and breaking bread, and it was real

we are bound by magic twists, and proof that they existed, now revealed

I'm rich with treasures in my chest

3. Web of Mystery

2018 ©Catalooch Music, BMI

Born of the realization that every one of those spaces on your family tree are essential to you being here, whether or not you know the name or anything else about the occupants of each space. That the gift of your life passed through all of them, and that they deserve to be remembered in some way, and hopefully someday, each of us will be too.

Jon Carroll - organ and accordion
Michael Rohrer - bass
Lisa Taylor - drums
AM - vocals and electric guitars

I've been trying to touch the past, I wanna know just why I am
reach back through the mists of time, so much I want to understand
You were a child once like me, full of wonder, full of pain
I know sometimes that you were scared, I know that sometimes, you were brave

> ah ha, ah ha ha, ah ha,
> this is the only life I've known
> In part you gave it to me
> look behind or look ahead, you are woven in the thread
> in my web of mystery

I'm still trying to touch the past, while the present slips away
I gotta know how your part goes, before those pages fade to grey
You were old once like me, I wish that I could see your face
no one else knows what's passed our eyes, hurt and anger, love and grace

Someday centuries from now, like coaxing ember into flame
since I've left a verse or 2 behind, someone might still speak my name
someone might still speak my name
someone might still speak my name

4. Left Behind

2019 ©Catalooch Music, BMI

The raucous and divisive rhetoric about people coming to these shores fleeing countries of origin for hope at something better is not new. Most of our ancestors left some similarly disparaged place to come here. Thus most all of us are not only descendants of immigrants, but our ancestors left behind family as well. Up until only a century or so ago, those who left might as well have died, for they might never be heard from again. To connect with the descendants of the "left behind" has been a powerful elixir indeed. But the concept of the leaving and the left behind is much more a part of the human condition than simply our physical movements.

Rachel Taylor - cello
Michael Rohrer - bass
Lisa Taylor - drums
AM - vocals and electric guitars
My people braved boats over angry seas, risked everything giving life to me

Left the home they'd known for a stranger's shore, hope and hunger for one chance more
Irish blessings at American wakes, glasses raised to the hearts that break,
gathered at the door to see them go, of the rest of their lives, we'll never know

hold on tight, storms on the ocean
hold on tight, storms on the ocean, hey
break through the doors slammed in your face, with the love that remained

> Left behind, left behind, I'll never let go in my heart and mind
> of all their prayers and desperate dreams
> Left behind, left behind, forever changed once you cross that line,
> the great divide that lies between, the leaving and the left behind

coyote howls in the Mayan night, a boy in the shadows slips from sight
the packet in his belly is his only chance, at someone else's promised land

put one foot down, in front of the other
one foot down, in front of the other, hey
climb the wall rising in your face, with the love that remained

A newborn face in the morning sun, another journey has just begun
for time is a river and the river flows, through a mother's love towards the letting go

don't turn back, even if you want to

can't turn back, even if you want to
the way ahead is yours to claim, with the love that remains

5. Passage/(Fathers Now) Our Meeting Is Over

2017 ©Catalooch Music, BMI/Traditional

The Lowden guitar that was I gifted by my Irish cousins during our 2017 visit is the gift that keeps on giving. Every time I pick it up it seems some new idea flows out. Passage was just a little instrumental noodling that hypnotized me, as did the traditional tune that I learned the previous summer in Maine from my friend Dick Scobie.

AM - vocal and guitar
Dustin Delage, Tony Denikos, John Rickard, Madeleine McKnight and Lisa Taylor - ensemble vocals

Fathers now our meeting is over, fathers we must part
and if I never see you any more, I'll love you in my heart
til we land on the shore, til we land upon the shore
til we land on the shore, and are safe for ever more

Mothers now...

6. Aretas Culver

2018 ©Catalooch Music, BMI

It was quite a five-year odyssey from "meeting" my 3-times great grandfather face to face, to helping dedicate a veteran's headstone befitting a volunteer in Abe Lincoln's army. From Antietam to Andersonville, the story of Aretas and his ill-fated townsmen of Bristol CT will always stay with me. I'd like to think he'd appreciate the heart behind all of it.

Rachel Taylor - cello
Jeff Arey - mandolin
Michael Rohrer - bass
Lisa Taylor - drums and harmony vocals
AM - lead and harmony vocals, guitars

My name is Aretas Culver, in summer of '62
I traded in my papers, for a uniform in blue
I took Abe Lincoln's bounty, and Union swore to save
what I've seen, since that day, haunts me to my grave

We were farmers, merchants, clockmakers, and I knew all their names
some escaping trouble, still others seeking fame
proud we marched down Main Street, while Bristol rang her bells
none could know our destiny, to glory or to hell

> Ohh-oh, I can't escape the dreams
> ohh-oh, don't you hear their screams
> ohh-oh, I will kill no more
> ohh-oh, let them fight their civil war

Put on a train to Maryland, with but a single drill
to Antietam's killing fields, upon that cornstalk hill
we faced their fiercest legion, in a withering hail of lead
in terror ran while rebel yells, echoed in our heads

We dug graves for many days, watched by lifeless eyes
then rifles marching at our backs, in hands that fought at our side
Little Mac called it victory, but for failure we were blamed
and if I might survive this war, I vow to clear my name

Four nights surrounded, siege at Plymouth town
til our white rag of surrender, stopped the shells from raining down
to Andersonville through gates of hell, its deadline we had crossed
when they paroled my skin and bones, my soul had long been lost

I am Aretas Culver, my breath is fading fast
tell my wife and children, I loved them to the last
the curse I leave this world with, before the fighting's done
is those that vote to start the wars, be those that face the guns

7. A Dram to the Holidays

I felt that the first song I wrote on the gift guitar should have something to do with those shared ancestors to whom we all owe our existence. It was right before a show I was doing that Christmas, and I started to wonder about the hard lives they would have led in mid-1800s County Down. Around the time of the winter solstice there the sun sinks dreadfully early in the afternoon, and the hard work of manual laborers probably didn't slow much in post-famine Ireland either. Despite the title, it really has little to do with another skill attributed to "my people"; the distillation of spirits other than of the holiday variety.

Rachel Taylor - cello
Michael Rohrer - bass
Lisa Taylor - drums and harmony vocals
Tony Denikos - harmony vocals
AM - vocals, acoustic and electric guitars

A flickering candle, on cold window glass, brave and small in the dark
it beams like a lighthouse, on winter shores, guide me safe home in its glow

The wee ones a'slumber, deep in their beds, tucked close in a room much too small
their dreams burn brightly, far more than they should, have reason enough to believe
Lord give them reason to believe

> Raise a dram to the holidays, long may they last
> where hopes for the future dance, with ghosts of the past
> hold tight to the best of what's been
> we're still here at the holidays again

They come home off the water, and home from the mines, with worry lines worn in their souls
outside the revelers, filling the square, voices raise hymns up on high
proclaiming that Christmas is nigh

The longest of nights, gives way to the dawn, its radiance christens the sea
there's sweets in the oven, a log on the fire, and an angel rests high on the tree
blessed angel keep watch over me

8. The Gift

When you go to a place you've never been and where your ancestors lived, and you meet living flesh and blood, and they send you back to America with their beautiful family guitar, made in the hometown of your ancestors, that is a life-changing moment worthy of more than one song. But this is the one dedicated to that precious gift, and all that comes with it.

AM - vocal and guitar

It sits here in my living room, on the stand with window view,
somehow the strings stretch out of sight, wrap softly round the words I write
I cannot tell you what it means, family ghosts and ancient dreams,
in moments where the veil is thin, I feel their presence beckoning

> Like a king returned his family sword
> Torn battle flag or coat of arms,
> and gold coin in a peasant's palm,
> All the things I might become,
> with this gift of family

A double helix crossed the sea, part of you the same as me
the proof is singing in my hands, magic sound from distant lands
On some far-off moonlit night, long after I have passed from sight
Will it still know these songs I weave, held in hands not yet conceived

The papers on my table, the lines that fill my head
a spirit-conjured melody, mystically surrounding me

> Like a king returned his father's sword
> torn battle flag or coat of arms,

gold coins in a peasant's hand,
all the things that I now am

9. Sons & Fathers

In my 25-year career, I have performed with my dad playing keyboards many times, but we have never recorded together. On an album dedicated to family history that also belongs to him, it was not only right but essential. So I tried to write something musically in his wheelhouse, and I'm happy that he liked it enough to be willing to do it. From the perspective of "the bridge" between aging parents and being a parent myself.

Warren McKnight - piano and organ
Michael Rohrer - bass
Lisa Taylor - drums
AM - vocals and electric guitar

We are all born screaming, naked and afraid,
looking back it's funny seeing, all those plans I made
We'd throw our balls and race our bikes, sometimes dare to dream
manhood ain't as hard as, fathers make it seem

I guess it's just the way of things, we take our place in line
like our fathers and their fathers did, when it was their time
Now we're slowly trading places, I suppose those are the rules,
father time won't suffer any fools

All those things we cannot say
perfectly imperfect at, getting out of our own way

but we do not, we will not, could not though we know it's true
Nothing else gets quite as tangled up in knots, as sons and fathers do

Leave behind settling old scores
Gotta meet him where he's at, on the bridge or that far shore

Now it's my turn in that easy chair, it's the hardest thing I've done
Though any man can take the gig, it ain't for everyone
the have tos and the must dos, credit and the blame we earn
Hoping what we teach, is what they need to learn

10. My Little Town

Every one of us in the seven generations I know of my father's father's line left home and started anew someplace else. My little village in Virginia is home to me in a way no place has ever quite been. A tribute to our ghosts and our perils too.

Jeff Arey - mandolin
Michael Rohrer - bass
Lisa Taylor - drums
Tony Denikos - harmony vocals
Stephanie Thompson - harmony vocals
AM - vocals, acoustic and electric guitars
We close our road, on the 4th of July, for the kids to ride by
with streamers on their bikes, in red and blue and white, let the judges decide
bring a dish to share, and your favorite folding chair, try to stay cool,
we read the famous words, just like we've heard, since we were kids in school

in my little town

Renamed for Honest Abe, somehow it became, where I put roots down
graveyard stones, hide more than bones, in the red clay ground
these sunken gravel roads, wind back in time and slowly disappear
the last dairy farm, keeps holding on, another year
in my little town

> When the fire breaks out, when the roof caves in
> when the snow's piled high, times are worst they've been,
> or progress of the present brings the future crashing down
> stand up with each other here in our little town
> try not to blink and miss, our little town

The holiday lights, cold winter nights, herald angels sing
from door to door, just like the year before, firelight caroling
we pray that it lasts, cherishing our past before it's gone
where neighbors still wave, stories we still save to pass on down
in my little town

a world of change swirls around my little town

11. Long Ago and Far Away

2019 ©Catalooch Music, BMI

My now 12-year old daughter engages in music in many different ways. I wrote this tune on her fiddle, which I don't play but she does still now and again. It was different for both of us - me letting her hold up my singing with her playing. It's raw and emotive for me, much like contemplating my ancestors and her descendants all within the same song. Our porch functions like a family room for the warm seasons of the year; I hope this feels like you're here with us on a typical summer night.

Madeleine McKnight - fiddle
AM – vocals

When I was just a child, full of hope and dreams
of everywhere I'd go, and everything I'd see
My grandmother's knee and stories that she'd saved,
of lives, that we'd lived, long ago and far away

In the face of my child, full of hope and dreams
of everywhere she'll go, and everything she'll see
I will keep treasures safe, and pass to her one day
twice the tales that were lived, long ago and far away

When I'll wear, shades of gray, around my wrinkled face
a child of my child, held in my embrace
in her turn she will share, stories that she saved
When it is I, who once lived, long ago and far away

12. Entrelazando

2019 ©Catalooch Music, BMI

My wife's family comes from Spain and eastern Europe, and has been much harder to trace than my own. But our kid is the only heir in the whole world to both Michelle's family stories and mine. My Cuban mother-in-law would appreciate that I titled this accordingly in Spanish, meaning "intertwining." Composed and performed on the beautiful Lopez handmade classical guitar from our dear friend Teresa, and dedicated to Queta and Madeleine.

AM - classical guitar

13. Anniversary (2,000 Years Ago)

2016 ©Catalooch Music, BMI

"On this day x number of years ago, y happened." Anniversaries are noted in news, on calendars and many other ways every day. And on this day long ago, some milestone good or bad happened for your ancestors and mine. Two millennia are too far back for me to visualize, so it's purely my imagination. And that somehow makes it feel quite real to me.

Rachel Taylor - cello
AM - vocals and dobro

Two thousand years ago today, a farmer walked his fields,
Wondered what the land would yield, before the winter snow
Two thousand years ago today, two young lovers kissed,
laughed and made a wish, beneath a shooting star glow

> Someday in a far and distant future
> Millennia beyond imagining
> as time ticks away, what will remain
> of how we lived and loved,
> today

Two thousand years ago today, anger hurled a stone,
to break another's bones and keep some fool upon a throne
Two thousand years ago today, a cog in a war machine,
drew his last breath on a distant plain, as he dreamed of home

Two thousand years ago today, a mama hushed her little ones,
kissed them on the head and tucked them into bed
turned her face towards the sky, for blessings from on high
prayed to guide them safe, on the unknown path ahead
she wondered,

14. Our Meeting Is Over (Reprise)

Traditional; arranged Lisa Taylor & Andrew McKnight

The voices of the ancestors come to life, a cappella. Arranged with my amazing friend and bandmate Lisa, and featuring some very dear people in my life too.

AM, John Rickard, Les Thompson, Tony Denikos, Dustin Delage, Madeleine McKnight, Stephanie Thompson and Lisa Taylor - ensemble vocals

Brothers now.our meeting is over, brothers we must part
and if I never see you any more, I'll love you in my heart
til we land on the shore, til we land upon the shore
til we land on the shore, and are safe for ever more

Sisters now....

15. When My Time Comes

©2019 Catalooch Music, BMI

Written for my friend Chris King's memorial service, held in our historic little chapel that he helped our congregation to buy some 20 years ago. Built by formerly enslaved families working by firelight in 1890, its wooden walls are soaked deeply with the hopes and despairs of people who lived through segregation and lynchings in the Jim Crow south. Chris had great reverence for them, and was committed to their story taking its proper place in our local history. He also was quite a scholar of the blues. I felt like an improvised slide piece of some sort would be a fitting sendoff, but on the morning of the service, the song wrote itself in fifteen minutes. I'll always consider it a treasured gift from my friend in the Great Beyond.

AM - vocals and slide guitar

When my time comes, don't shed a tear (repeat)
I've got all I need and it's beautiful here

When my time comes, raise your voice & sing (repeat)
it sounds like angels when, that joyful noise rings

When my time comes, hold out your hand (repeat)
don't worry about me, lift up your fellow man and woman too

When my time comes, yeah ring them bells (repeat)
until we meet again, fare thee well

APPENDIX C
Finding Your Stories

The goal of any big creative project is in some way to touch your audience. My hope is that you have been inspired by some of what you have read and heard in this project to dive more deeply into your own ancestories, for yours are just as precious as mine. I've done my best to collect and curate some of the best wisdom to which I've been exposed, to hopefully save you some time and frustration. To that end, Appendices B and C are intended to be quick reference guides you can return to many times.

In a sense, helping others to become better family historians helps all of us, for we never know where we might find a new thread connecting us to new people. This book is full of examples of how easily those moments can come from out of the blue. I never stop learning new techniques for finding and analyzing new evidence, especially as more old records are digitized and made available online. Literally, we will never know less than we do right now, and it is exciting to think of what we might learn over the next year and decade.

Film Your Elders

My first, most important and vital piece of advice is this. Our elderly relatives are closer to the past than we ever will be. Whenever it's possible, and whenever the source can be considered at least moderately reliable, film them telling their stories. "Tell me about your childhood, where were you born and when? What do you remember about growing up?" As a researcher, you can always return to these recorded sources.

As a human being, it is hard to overstate the value of being able to see and hear them again long after they have departed this earthly plane. For that sentimental value alone, it is hugely worth it. And consider the generation or two after you - to able to see and hear their great-great grandmother in person is something that you most likely never got to do. Can you imagine from where we sit now what that would be like?

Save old stuff online and back up media

My second big takeaway - save your work in multiple places that are truly safe. Those precious old pictures and letters need to be scanned to the best quality you can manage, and saved to a safe and secure online location. Not only does this mitigate the disaster a bit if there is a fire or flood, but it allows you to selectively share and collaborate with other living relatives. We each add our pieces to the puzzle.

Write Down the Necessities to Review and Interpret Your Work

Third, write it down and save that someplace too. If you were to die suddenly, would your family lose access to all that you've learned? Would they know what you have accumulated and how to get it? Would they understand any of what you've done, without you there to explain it to them? And are your passwords available someplace so some interested party can actually revisit and perhaps continue your hard work?

Use the tools of the 21st century to make it easier. Talk-to-text resources allow us to quickly dictate important information for later editing. Many of these tools on modern smart phones automatically back up and sync online. I use Apple Notes, Evernote and a simple voice memo recorder as essential parts of my work - I create something on my phone, and it is waiting for me on my laptop. As long as someone can access your information someday - preferably with your blessing - it would be possible for them to review your notes, even if you never did anything more with them.

Take good notes, and do your best to identify the sources for things you've learned. When you share your opinions, make it clear that that is what you are doing. Good theories can be useful to future work, even if you never get anywhere proving them now. You may (and should) collaborate with others here in the present, or by handing your work down to some future researcher.

Your ancestors were real people just like you and I, and their stories truly belong to all of their descendants. And consider your legacy - what you've done to help bring them to life should be preserved and passed on so that others may know them too.

DNA Testing

Fourth, test your DNA and that of any willing elderly relatives, and upload your results to the other big testing companies to cast your cousin net as widely as possible. Study the privacy issues, accept what you are willing and avoid the rest. DNA alone doesn't prove anything in your genealogy, but when used properly with full knowledge of its limitations, it can provide solid evidence to further support your paper trail. At a minimum, get some close cousins on both sides of your family to test, preferably ones that are known close relations to only one of your parents. (No disrespect intended to those like my wife's set of double first cousins, thanks to two Polish brothers marrying two Cuban sisters!)

Appendix D addresses using DNA matching to advance your family research.

Going Forward - Build a Tree

Now that those four big foundational points are addressed, the starting point is to create a tree that includes all that we know. Especially important information includes birth, marriage and death records

(BMD) including verified dates and locations. The wonderful free resources at FamilySearch.org can be a fantastic asset in getting started with confidence. You can also build your tree there, and even export it to another website or software. (Note: family trees are stored as GEDCOM files and have a standard file format just like images, movies and documents - extension .ged.)

To make the most of your work, consider my "successive umbrella" strategy for identifying levels of cousins. Add your parents' siblings and their descendants to your tree. Include the second marriages and half-cousins too. When you've finished this to the best of available knowledge, move on to the siblings of each of your four grandparents - the grandchildren of those siblings will be your second cousins (2Cs) and so on up to the next level, the descendants of your eight great-grandparents' siblings. Your common ancestors with them will be two or perhaps one of your great-great (2G) grandparents. You may find that you have hundreds of 3Cs, but even if you don't know all of the living descendants, you will have a valuable database of surnames associated with each set of 2G grandparents.

Why is building a tree so important? It is extremely likely that some of those hundreds of cousins may be researching a branch of their family that you have in common. You may learn a lot simply by being "discoverable" and connecting with other family historians. This becomes even more important if you have done a DNA test, for there are millions of people using this powerful tool in their family research.

Take Care with Clues - Evidence, Not Fact

Before we dive deeply into the past, we need a couple of really important tools at all times. It is exciting to start down the trail into the mists of the past to explore our family origins. We may have been told some fascinating story about great-grandpa and a harrowing escape during the war, that we have Native American heritage, or an adoptee may have heard that his birth mother lived in a small town outside of Houston and moved to Winnipeg. It is human nature to seek evidence for a belief or a story that we may have held onto (like the one in my family about being descendants of Robin Hood).

These are important clues to research! But, they are not facts until they have been proven. If we go into our research seeking to validate a story, it can taint how we interpret the evidence - to raise its importance if it supports our belief, or discount it when it doesn't. We must take great care not to let preconceptions color our research - particularly when the clues in the records and/or DNA evidence might add up to a different, conflicting narrative.

Timelines are extremely useful tools, for they can give us lots of clues about the fate of our ancestors, their mobility, and what large factors like war, epidemic and famine may have been important in their lives. I have a handy timeline of big inventions, such as the advent of rail and plane travel, and events of global significance such as the World Wars. I have timelines on the same spreadsheet for the countries or regions of countries where my ancestors lived, to make it easier to evaluate how those events might have affected them.

Historical records are fraught with fuzziness. Spelling of surnames in particular was not a precise requirement even a century ago. Census takers wrote down the information they were told, including ages and names, and often there are inconsistencies to be found for a single person over two or three censuses. Locales change hands as a result of invasions and independence movements, which may affect where records are stored for certain time periods and even the language in which they were recorded. Endogamy - multiple marriages between several families - is historically quite typical in small communities without much influence from migration, and sometimes BMD records can be confusing as to which John Ryan they refer. Lifespans were considerably shorter two centuries ago, and so widowed people frequently married again, often while still of child-bearing age. Even gravestones are etched with inaccuracies of date and place.

An Open Mind is the Sharpest Tool

An open mind is essential in interpreting our findings. So too is being thorough and complete when analyzing evidence. For example, to look at all of the information included in the census - not just the records of our family members and any boarders living with them, but who lived next door and on the same street. What work did they do? How much schooling did they have? Where were their parents born? And, did the machine - or the census-taking human with the pen - transcribe everything correctly? Variations in spelling can mean a missed hint, and in olden days it seems that spelling in general could be a much looser business!

Never before have we had so many powerful tools, from cheap DNA test kits to literally tens of millions of digitized old records in many countries, easily searchable on the Internet, and the character recognition technology to do so. You can learn more about your family history from the comfort of your own home than has ever been previously possible. And of course it is also possible to make more mistaken assumptions as well.

A Reality Check

It is important to realize that this work is never truly complete. It is also true that the dead aren't going anywhere. While your trial subscription to a valuable website may run out soon, most deadlines are self-imposed. It is easy to get dragged willy-nilly through the online catacombs containing your ancestors' records, bouncing from one to another without purpose or plan. Having a strategy and a goal for each session, and keeping at least a semblance of a log as you work, will help maintain your sanity and improve the quality of your work.

Collaborating with others will help you stay focused on a specific shared ancestor or family. It is wonderful to learn things from each other, and it helps ensure that the ancestory will be told and preserved. While I have built my family tree at Ancestry (and saved it at home through Family Tree Maker software) and it is easily exportable, I have made my family tree at the collaborative site WikiTree one person at a time. WikiTree's lofty purpose is to build one genetic family tree that connects all of us. I use that time to make sure I have my information for that ancestor as accurate as possible, and any theories or fuzzy evidence is duly noted. It is a check on my work, but also an effort to make sure I don't degrade the known facts about any of my ancestors for others who might be researching them now or in the future.

As hard as it might be to imagine, you will be ancient history one day too. Don't ignore capturing important moments in your story as you go. Someday your descendants may be very grateful you did.

APPENDIX D
More About DNA Testing

The most common DNA tests done by tens of millions of people evaluate our autosomal DNA. These tests provide evidence that can help to identify and confirm ancestors born in the early to mid 1800s, a time when geographic mobility was generally much less than today. Chances are if you can trace your family back to a specific region at that time, your ancestors may have been there for a few generations. Learning more about the history of that place may well help discover other clues.

The basic goal of genetic genealogy is to collect evidence to validate a researched connection to a common ancestor. We do this by evaluating a person who matches our DNA with our shared matches – the people with whom we also share those same specific bits. For adoptees and others searching for their birth parents, DNA matching provides solid and indisputable evidence of recent ancestry.

Matching DNA amounts are quantified usually in centiMorgans (cM), or sometimes as a percentage of shared DNA. While DNA matching is a powerful tool that can contribute a wealth of invaluable evidence to our research, it is vital to understand its value and its limitations clearly:

- By itself, you can't tell if a match is maternal or paternal; this can only be determined by comparison to known cousins.

- DNA doesn't always point to a simple answer. Sometimes things happen, not all of them good or pleasant. There are lots of logical explanations when the DNA evidence (or inexplicable absence of it) does not support the family tree or history that's been handed down, and usually it means that something was handed down that differed from reality. Consider all of those possibilities carefully, and gently.

- It is exceedingly rare to not share significant DNA with second cousins and closer relationships. The absence of an appropriate amount of shared DNA with your close kin is evidence of a misattributed parentage somewhere in that branch of the family.

- While the practical threshold for an actual match varies typically between 7 and 10 cM (depending on which expert you ask) it is usually difficult to identify a common ancestor with matches of less than 20 cM. (Once in awhile you get lucky and both people have well-researched trees that reach far enough back in the common branch.)

- We also can't determine which person - you or your match - is closer to the common ancestor from the DNA amount alone.

- Some segments of DNA pass through many generations unbroken, and originate too far into the past to be able to identify a common ancestor.

- The different testing companies use different algorithms to evaluate your DNA, so the amount you show in common with the same DNA match can vary considerably from one company to the next.

- DNA doesn't "skip a generation." You inherited all of your DNA from your two parents. In my case, I occasionally get small matches with people who don't match either of my parents, and thus I assume that they are false positives produced by that company's matching algorithms.

- Lots of other relationships are possible - half relations for instance. And it wasn't that long ago that many people lived in small communities and routinely intermarried amongst families (2nd cousins marrying was quite common in the early 1800s). The practice of marrying exclusively within one's own ethnic, cultural, social, religious or tribal community is known as endogamy, and it was common in colonial America and just about everywhere else too. If you have colonial ancestry, it is quite possible to share several distant common ancestors with a single match.

The more testing we do with more known points of origin, the more we will learn about migrations of people over the last millennia. With the growth of that DNA database, we learn more about the legacy of the millions of people stolen out of Africa until the mid-1800s, and countless other peoples whose stories have largely been lost. As with many topics in genealogy, one could easily say we will never know less than we know right now.

Consider the immensity of life extinguished in Europe in 1914-18, and around the world in the Second World War. Each day there are people doing the arduous forensic detective work when remains are uncovered. The millions of people who have already done DNA tests make identifying these unknown dead more possible.

As temperatures warm above freezing more of the year in northern latitudes, the earth is yielding remains of people who lived hundreds or thousands of years ago, in places and periods of history that we understand more anecdotally than scientifically. Because of the large and growing knowledge base about genetic signatures from different ancient communities, we not only have a better understanding of how people lived in a place, but are also revisiting and refining our theories about how and why they moved from one place to another. Other advances are now allowing us to test artifacts and long held family treasures like letters and locks of hair.

People quite understandably and prudently have large concerns about privacy. One doesn't need a vivid imagination to consider how DNA could be used for inconvenient and nefarious purposes - insurance companies and fringe medical researchers as well as sophisticated criminal enterprises could certainly find ways to improve their business models and better target their markets.

However, it is inevitable that DNA testing will continue to bring progress and benefit in our everyday lives too. DNA testing for genealogy is not a profitable venture, even as the databases of large companies get into tens of millions. In order to continue five or ten years from now, these companies whose profitability is essential to retaining their skilled employees as well as our collective genetic knowledge base will need to continue to find ways to make money. It is good for us to work like that future is not inevitable, and to save and document our DNA work to the best of our ability. It is essential to not risk losing all that knowledge if someday the genealogical division of an Ancestry or 23andMe goes dark.

There are exciting new DNA tools and techniques being developed for genealogy by talented individuals with programming skills as well as companies. New techniques such as match and tree clustering, and chromosome mapping offer new chisels to put to our genealogical brick walls.

A word of caution about the living is important too. Not everyone in your family will share your excitement about everything you learn. We routinely find that what is acceptable behavior in our modern world was scandalous and a source of shame during the youth of some of our elderly relatives. There will be stories and secrets, and the potential for deeply hurt feelings should always be considered.

We are all human after all, and the DNA will certainly confirm that. This means that we consider the whole life - the good, the bad, and all the between. To do so with impartiality and sensitivity is as important as doing careful and thorough research.

So, onward then. With curiosity and excitement, tempered with a good dose of human decency!

APPENDIX E
Actual DNA Relationship Ranges

We can predict the amount of DNA we should share with our relatives based simply on the amount of DNA we would expect to inherit from each generation, but in practice we find that the more distant the relationship, the more widely the results vary from those predictions. Like any sound science, rigorous data collection and statistical analysis help better interpret our actual results. For much more about the science of interpreting results, the International Society of Genetic Genealogy Wiki page is invaluable https://isogg.org/wiki/Autosomal_DNA_statistics. Dr. Blaine Bettinger has done an extensive crowdsourced research project on actual matches with known relations in the Shared cM Project (see next page), which shows the range of minimum, maximum and mean values for each relationship.

The Shared cM Project – Version 3.0

August 2017

For MUCH more information (including histograms and company breakdowns) see: goo.gl/Z1EcJQ

Blaine T. Bettinger
www.TheGeneticGenealogist.com

How to read this chart:

Aunt/Uncle	← Relationship
1750	← Average
1349 - 2175	← Range (low-high) (99% Percentile)

								Great-Great-Great-Grandparent		GGGG-Aunt/Uncle	
							Great-Great-Grandparent		GGG-Aunt/Uncle		
Half GG-Aunt/Uncle 187 12 – 383	Great-Grandparent 881 464 – 1486							Great-Great Aunt/Uncle 427 191 – 885			Other Relationships
	Half Great-Aunt/Uncle 432 125 – 765	Grandparent 1766 1156 – 2311					Great Aunt/Uncle 914 251 – 2108				6C 21 0 – 86
		Half Aunt/Uncle 891 500 – 1446	Parent 3487 3330 – 3720			Aunt/Uncle 1750 1349 - 2175					6C1R 16 0 – 72
Half 3c 61 0 – 178	Half 2c 117 9 – 397	Half 1C 457 137 – 856	Half-Sibling 1783 1317 – 2312	Sibling 2629 2209 – 3384	SELF	1C 874 553 – 1225	2c 233 46 – 515	3c 74 0 – 217	4c 35 0 – 127	5c 25 0 – 94	6C2R 17 0 – 75
Half 3c1R 42 0 – 165	Half 2c1R 73 0 – 341	Half 1C1R 226 57 – 530	Half Niece/Nephew 891 500 – 1446	Niece/Nephew 1750 1349 - 2175	Child 3487 3330 – 3720	1C1R 439 141 – 851	2c1R 123 0 – 316	3C1R 48 0 – 173	4C1R 28 0 – 117	5C1R 21 0 – 79	7C 13 0 – 57
Half 3c2R 34 0 – 96	Half 2c2R 61 0 – 353	Half 1C2R 145 37 – 360	Half Great Niece/Nephew 432 125 – 765	Great-Niece/Nephew 910 251 – 2108	Grandchild 1766 1156 – 2311	1C2R 229 43 – 531	2c2R 74 0– 261	3C2R 35 0 – 116	4C2R 22 0 – 109	5C2R 17 0 – 43	7C1R 13 0 – 53
Half 3c3R	Half 2c3R	Half 1C3R 87 0 – 191	Half GG Niece/Nephew 187 12 – 383	Great-Great-Niece/Nephew 427 191 – 885	Great-Grandchild 881 464 – 1486	1C3R 123 0 – 283	2c3R 57 0 – 139	3C3R 22 0 – 69	4C3R 29 0 – 82	5C3R 11 0 – 44	8C 12 0 – 50

Minimum was automatically set to 0 cM for relationships more distant than Half 2C, and averages were determined only for submissions in which DNA was shared

About the Author

While singer, writer and guitarist Andrew McKnight has been a prolific songwriter, performer and recording artist for a quarter century, *Treasures in My Chest* is his first full-length book. A former engineer with advanced degrees in Environmental Engineering and Chemistry, he turned his scientist's eye and songwriting skills towards his family history. He has released nine recordings, and performed at many prestigious venues like the John F. Kennedy Center, the International Storytelling Center and the Atlanta Olympics. Andrew has used his writing skills to write songs with kindergarten kids and wounded warriors coping with post-traumatic stress. He is a gifted workshop leader, an engaging speaker, and a longtime endorsing artist for Elixir Strings and Fairbuilt Guitars. Mr. McKnight lives in Lincoln, Virginia with his wife, daughter and a considerable cadre of friendly ghosts.

About the Author